PHYSIOLOGICAL DISORDERS IN HORTICULTURAL CROPS

PHYSIOLOGICAL DISORDERS IN HORTICULTURAL CROPS

K. Balakrishnan, Professor of Crop Physiology
Agricultural College and Research Institute, TNAU, Madurai-625 104

J. Rajangam, Professor and Head
Department of Fruit Crops
Horticultural College and Research Institute, TNAU, Periyakulam-625 604

A. Subbiah, Asst. Professor of Horticulture
Grapes Research Station, TNAU, Theni, Tamil Nadu

S. Parthiban, Professor and Head
Grapes Research Station, TNAU, Theni, Tamil Nadu

P. Balasubramanian, Asst. Professor of Horticulture
Department of Horticulture
Agricultural College and Research Institute, TNAU, Madurai-625 104

M. Prakash, Professor
Department of Genetics and Plant Breeding,
Faculty of Agriculture, Annamalai University, Chidambaram-608 003

SSPH

SATISH SERIAL PUBLISHING HOUSE
403, Express Tower, Commercial Complex
Azadpur, Delhi-110033 (India)
Phone : 011-27672852, Fax : 91-11-27672046
e-mail : info@satishserial.com, hkjain1975@yahoo.com
Website : www.satishserial.com

Published by :

SATISH SERIAL PUBLISHING HOUSE
403, Express Tower, Commercial Complex, Azadpur, Delhi-110033 (INDIA)
Phone : 011-27672852 Fax : 91-11-27672046
E-mail : info@satishserial.com, hkjain1975@yahoo.com

ISBN 978-93-86200-89-1

Composed, Designed & Printed in India

ABOUT THE AUTHORS

Dr. K. Balakrishnan is presently working as a Professor of Crop Physiology at Agricultural College and Research Institute, Madurai. He was the recipient of ICAR Senior Fellowship during 1983 and obtained Ph.D in Crop Physiology in 1987 from Tamil Nadu Agricultural University, Coimbatore (TNAU). He has also obtained Post Graduate Diploma in Journalism in 1988. He also worked in United Planter's Association of South India, Coonoor, Nilgris, Tamil Nadu. He has been teaching courses on Crop Physiology and Biochemistry to the Under Graduate and Post Graduate students in both Agriculture and Horticulture for the past 32 years.

He served as Principle Investigator of Foliar diagnosis in Horticultural crops in plan scheme funded by Government of Tamil Nadu, Post Harvest Technology of Horticultural crops ICAR-PHT at Horticultural College and Research Institute, Periyakulam and Arid zone fruit crops funded by AICRIP (ICAR), at Regional Agricultural Research Station, Aruppukkottai. He operated several schemes from both private and Government agencies. He also worked as Assistant plant physiologist in ICAR-Tropical fruit crops, AICRIP, Horticultural College, Coimbatore. He was the member of All India ICAR Syllabus Committee on basic Agricultural sciences during 2007-2008. He also served as a member of Board of studies, Tamil Nadu Agricultural University during 1997-2000.

His expertise in foliar diagnosis of nutritional disorders in crop plants is well recognized by invitation to deliver lectures on various aspects of diagnosis of nutritional and physiological disorders in NGOs, Extension officials and farmer's organisations. He has solved many physiological field problems. He has authored and co-authored 10 book chapters and 6 books and published 60 research papers in both national and International journals, 3 extension bulletins and 120 popular articles. He has been the question setter and external examiner for M.Sc.(Ag.) and Ph.D thesis in several Universities. He has organized one national seminar at AC&RI, Madurai during September 2013. He served as the Professor and Head, Department of Seed Science and Technology, Agricultural College and Research Institute, Madurai during 2013-2015.

Recognition of his 25 years of meritorious service, TNAU award given on 41st foundation day celebration held on June 01, 2011. He was awarded Thiruvalluvar Award for his contribution in Teaching Crop Physiology courses to the under graduate students in Agriculture by Alumni Association of Agricultural College and Research Institute, Killikulam in 2017. He has been awarded Fellow of Bose Science Society (FBSS) by the Tamil Nadu scientific Research Organization, Tamil Nadu during January 2018.

Dr. J. Rajangam, started his career as an Asst. Professor (Horticulture) in the year 1995 at Tamil Nadu Agricultural University (TNAU), Coimbatore. As a Horticulturist, he has more than 20 years experience in various horticultural crops. During his service, significant contributions were made pertaining to release of varieties CO 3 Chillies, PKM 2 Annual moringa and PKM 1 Davana. He served as Professor and Head, Horticultural Research Station, Kodaikanal for a period of four years. During his tenure, he was officer-in-charge of All India Network Project on Temperate Fruit Crops at Horticultural Research Station, Kodaikanal with CITH, Srinagar and PI of ICAR - AICRP (Fruits) at Horticultural College & Research Institute, Periyakulam. He got rich experience is teaching UG and PG courses for more than 16 years and guided 12 M. Sc (Hort.) students. He has published more than 30 papers in National and International journals, 112 popular articles, 15 booklets and two books. He has visited Israel and Tanzania for Post Graduate Certificate Programme in Protected Horticulture and International Symposium on Moringa cultivation and utilization. Currently, he is serving as the Professor and Head, Department of Fruits, Horticultural College and Research Institute, Periyakulam (TNAU), Coimbatore.

Dr. A. Subbiah is presently working as an Assistant Professor of Horticulture at Grapes Research Station, (TNAU), Theni, Tamil Nadu. He was the recipient of ICAR Junior Research Fellowship for his Master Degree during 1997 and completed his Ph.D., in 2004 from Tamil Nadu Agricultural University, Coimbatore. He also worked as Research Assistant at Cardamom Research Station, Pampadumpara, Idukki, Kerala (KAU), Senior Research Fellow at Department of Vegetable Crops, TNAU, Coimbatore and Technical Executive of Indian Spices Producers Company, Cochin, Kerala. He has also been involved in teaching courses on fundamentals of horticulture, production and postharvest technology of fruits to the Under Graduate students in Horticulture for the past 10 years. He also served as Co-Principle Investigator in the ICAR-NICRA Research Project on "Studies on Mango Flowering Phenology" and IDRC-CIFSF, Canada funded Project on "Enhanced Preservation of Fruits in Asia by using Nano Film Techniques". Further, he is also In-Charge of Venture Capital Scheme "Mass multiplication of grape rootstocks and varieties. He was associated with 10 various scientific societies involved in promotion of research and development in the field of fruits, spices and plantation crops as life member. He also served as a member of Board of Studies for UG Programme in Horticulture, Tamil Nadu Agricultural University during 2010-2013. He has authored and co-authored 4 book chapters and 6 books and published 23 research papers in reputed National and International Journals. He has been the question setter for B.Sc. (Hort.) Programme of UHS, Bagalkot. He was the organizer of various training programme related to production technology of mango, grapes and banana for the benefit of farming community. He is currently involved in research activities related crop improvement and canopy management in grapes.

Dr. S. Parthiban is working as Professor and Head, Grapes Research Station, Royappanpatty, Theni District, Tamil Nadu. He has assembled 120 grapes varieties and distributed 2.0 lakhs Dogridge rootstocks to grapes growers. He worked in Coffee Board scheme and ICAR-AICRP on Floriculture Scheme at Horticultural Research Station, Yercaud. He was the Officer-in-charge for the scheme ICAR-AICRP on Tropical Fruits at Horticultural College and Research Institute, Periyakulam for 8 years and specialized in research on sapota, acid lime and jackfruit. He was one of the scientists responsible for the release of new crop varieties *viz.*, PKM-4 and PKM-5 sapota, PKM-1 Manila tamarind, PKM-1 Dharna and two technologies in mango. He also worked as Professor and Head at Horticultural Research Station, Thadiyankudisai and carried out research on coffee, hill banana, avocado, mandarin orange, pepper, cinnamon, nutmeg and litchi. He underwent 15 days training programme on "Experiential Learning" at University of Western Sydney, Australia during 1996. He served as member, Board of Studies, Tamil Nadu Agricultural University during 1996-2001. He was the approved question paper setter and evaluator for B.Sc. (Ag.) & M.Sc. (Ag.) courses in Annamalai University, Agri-UG & PG diploma courses in Gandhigram Deemed University and approved external examiner at Dharwad Agricutural University. He also served as NSS officer for three years, Professor (Farms) for seven years and Professor and Head, Department of Vegetable Crops at Horticultural College and Research Institute, Periyakulam. He has 16 years of teaching experience and handled several courses for B.Sc. (Hort.) and M.Sc., (Hort.) students. He has guided 15 M.Sc. (Hort.) and 4 Ph.D. (Hort.) students. He has authored 14 book chapters and 7 books and published 42 research papers.

Dr. P. Balasubramanian started his career as an Assistant Professor in Horticulture, in the year 2009 at TNAU, Coimbatore. As a horticulturist he has more than 9 years Research experience in various horticultural crops. During his service, significant contributions were made pertaining to the release of varieties MDU-1 cluster beans. He served as a Farm Manager at Soil and Water Management Research Institute, Thanjavur and also at the department of Horticulture, Agricultural College and Research Institute, Madurai for a period of 7 years. He is a principal investigator of BRNS-GOI, Mumbai scheme entitled "Isolation of short statured early maturing mutant in cluster bean through gamma irradiation (*Cyamposis tetragonoloba* L.") with a budget of 24.11 Lakhs. He is offering courses in Horticulture for more than 7 years for U.G, P.G, Ph.D students in Agricultural and Horticulture and also guided 2 M.Sc. and 1 Ph.D. He has published 13 Research articles in peer reviewed national and international journals.

Professor M. Prakash born in 1966, obtained his B.Sc.(Ag.) and M.Sc.(Ag.) from Tamil Nadu Agricultural University, Coimbatore, India in 1988 and 1990 respectively. He did his Ph.D from Annamalai University. His studious and scholarly aptitude during his studenthood fetched him prestigious student support awards like National Merit Scholarship, Government of India during 1981-1988 and ICAR Junior Fellowship during 1988-1990. Prof. M. Prakash is working on abiotic stress tolerance from his post graduation onwards. He started working on effect of antitranspirants on growth and physiology of brinjal and continuing his work on abiotic stress tolerance till date. He is having 25 years of teaching and research experience in the Department of Genetics and Plant Breeding wherein he has guided 8 Ph. D. and 22 Masters Degree Programmes. He has also completed five funded research projects.

He is presently serving as a Professor, Department of Genetics and Plant Breeding, Faculty of Agriculture, Annamalai University. He has earned the status of Departmental Research Support with Special Assistance Programme of University Grants Commission (UGC-SAP- DRS) to the Department with financial assistance of Rs. 50 lakhs at Phase I (2009-2014) and Rs. 1.025 Crore at phase II (2015-2020) level.

His credentials include publications of 85 referred research papers, seven books published by National publishers, three books published by International publishers and one book chapter published in SESAME by globally reputed CRC Press. He has significantly contributed in knowledge sharing, through organisation of 7 National Seminars, participation in 8 International Seminars, Conferences and Workshops and 18 National Seminars and Conferences. His expertise in abiotic stress tolerance is well recognised by invitations to serve in the Editorial Board / Committees of several academic journals, awards like Crop Research Award in 2004 and Best Researcher Award in Annamalai University in 2010 and Fellowships like FELLOW of Indian Society of Plant Physiology, New Delhi in 2015 (FISPP), FELLOW of Association for the Advancement of Biodiversity Science, Bangalore in 2015 (FABSC) and FELLOW of National Academy of Biological Sciences (NABS), Chennai in 2016(FNABS).

Preface

High technology and high input philosophy are the modern concept to maximize the productivity to cope up with the needs on food, fuel, fodder and fibre of exploring population in India. Nutrient mining by intensive cultivation has made the soil malnutritioned leading to expression of nutrient deficiencies and physiological disorders in crops which ultimately resulted in reduced output

Physiological disorders are abnormalities in crop growth that occurs in plant organs caused due to improper physiological functioning of the plant. These disorders in crops are mainly due to sudden drift in environmental factors, low soil fertility, poor quality of irrigation water, pest infestation, misuse of chemicals such as herbicides, growth regulators, pesticide, etc. Physiological disorders are more common in horticultural crops than in field crops due to their inherent biochemical complexities. Therefore, better understanding of the nature, cause and remedial measures of physiological disorders in horticultural crops, a book on "Physiological disorders in horticultural crops" has been brought out.

This book comprises of cause, symptoms and remedial measures of different physiological disorders in fruits, vegetables, flowers, spices and plantation crops. In addition, some useful information related to plant nutrition and physiological disorders are also given in appendix.

In this book, I do not pretend to claim credit for any original contribution on the subject. Rather, it should be viewed as a concise collection of various topics from different books, bulletins, journals, scientific paper, etc., the names of which are indicated in the text itself or in the selected reference. I have tried my best to meet the greater needs of the students of horticulture of different Agricultural and Horticultural universities as well as extension workers across the country. I express my thanks to Tmt. S. Vijiganapathy of this department for neat typing this manuscript.

K. Balakrishnan
J. Rajangam
A. Subbiah
S. Parthiban
P. Balasubramanian
M. Prakash

Contents

I. Introduction

Physiological disorder is defined as the abnormalities of abnormal growth that occurs in the plant organs caused due to improper physiological functioning of the plant. These disorders in crops are mainly due to sudden drift in the environmental factors, changes in the fertilizer pattern, quality of irrigation water, pest infestation, misuse of chemicals such as herbicides, growth regulators, pesticides, etc. These factors often interact or interplay in producing an effect. It is a non-infectious disorder and some are easy to diagnose while others are difficult or even impossible to diagnose. Some non- infectious disorders are beyond help once symptoms are visible.

Guidelines to diagnose physiological disorders

- Examine where symptoms first appear on a plant—on new leaves, old leaves or all over?
- Note the pattern of any discolouration or yellowing—is it all over or between the veins or around the edges?
- Note general patterns rather than looking at individual plants—are the symptoms distributed throughout a group of plants of the same type growing together ?. In the case of a deficiency, all the plants should be similarly affected, although the distribution will depend on past treatments applied to the soil.
- Soil analysis, such as determining pH, can help to confirm the presence of physiological disorders.
- Consider recent conditions, such as heavy rains, dry spells, frosts, etc., which may also help to determine the cause of plant disorders.

Factors affecting physiological disorders

a) Soil Compaction

Soil structure determines the soil's ability to hold water, nutrients and oxygen and make them available to plants. The most common issue related to soil structure is compaction, which results in inadequate pore space for root growth. Clay soils with their smaller particle size, have naturally smaller pore space and are at high risk for becoming severely compacted. Compaction can occur from a variety of sources including traffic (particularly heavy farming or construction equipment), raindrop impact, tilling operations and minimal crop rotation. Reduced water availability is an obvious consequence of compaction as runoff occurs more frequently in a compacted soil and available pore space to hold water is limited. However, low oxygen availability for root respiration can also be a serious consequence of restricted soil pore spaces.

b) Soil pH

Soil pH is the measure of the H^+ ion activity in the soil solution. A high amount of H^+ activity results in an acidic soil condition, while low activity results in a predominance of OH^- activity, leading to alkaline soil. Although some plant species have preferences for more extreme acidic or alkaline soil conditions, it is generally regarded that a slightly acidic pH range of 6-7 is most favourable for plant growth. Soil pH outside of this range can have a dramatic impact on the solubility and therefore, availability of plant nutrients. Soil pH below 5.5 generally results in low availability of calcium (Ca), magnesium (Mg) and phosphorus (P) and increased solubility of aluminum (Al), iron (Fe) and boron (B). High level of these three nutrients in low soil pH are common and can induce toxicity symptoms in plants. Soils with pH levels above 7.8 have a high availability of Ca and Mg at the expense of P, B, Fe, Mn, Zn and Cu. Plants grown in these alkaline soils often have deficiency symptoms to these nutrients.

c) Nutrient Deficiencies

Nutrient deficiencies often result from a lack of plant nutrients in the soil. Two of the easiest ways to recognize nutrient deficiencies are the lack of visible pathogen signs (infectious microbe parts, such as mycelium) and the relatively uniform distribution pattern of symptoms in the field as compared to many diseases caused by plant pathogens. However, plant nutrient deficiencies are best diagnosed using plant tissue analysis. As opposed to soil nutrient analysis, plant tissue analysis allows one to determine plant nutrient uptake rather than plant nutrient availability. Because nutrient deficiencies lack visible signs, they are often mistaken for virus diseases. One of the best ways to diagnose nutrient disorders is the distribution of symptoms on the plant. Mobile nutrients are readily transferred within the plant to the growing points, so symptoms appear on the lower (older) leaves of the plant. Conversely, immobile nutrients display symptoms on the meristem of the plant during nutrient deficiencies.

Deficiency symptoms

Nitrogen

- Stunted growth may occur because of reduction in cell division.
- Pale green to light yellow colour (chlorosis) appearing first on older leaves, usually starting at the tips. Depending on the severity of deficiency, the chlorosis could result in the death and/or dropping of the older leaves. This is caused by the translocation of N from the older to the younger tissues.
- Reduced N lowers the protein content of seeds and vegetative parts. In severe cases, flowering is greatly reduced.
- Nitrogen deficiency causes early maturity in some crops which results in a significant reduction in yield and quality.

Phosphorus

- Due to phosphorus deficiency, the initial overall symptom is slow, weak and stunted growth.
- Phosphorus is relatively mobile in plants and can be transferred to sites of new growth, causing symptoms of dark to blue-green colouration to appear on older leaves of some plants. Under severe deficiency, purpling of leaves and stems may appear.
- Lack of P can cause delayed maturity and poor seed and fruit development.

Potassium

- The most common symptom is chlorosis along the edges of leaves (leaf margin scorching). This occurs first in older leaves, because K is very mobile in the plant.
- Because K is needed in photosynthesis and the synthesis of proteins, plants lacking K will have slow and stunted growth.
- In some crops, stems are weak and lodging is common if K is deficient.
- The size of seeds and fruits and the quantity of their production is reduced.

Calcium

- Calcium is not mobile and is not translocated in the plant, so symptoms first appear on the younger leaves and leaf tips. The growing tips of roots and leaves turn brown and die.
- Newly emerging leaves may stick together at the margins, which causes tearing as the leaves expand and unfurl. This may also cause the stem structure to be weakened.
- In some crops, younger leaves may be cupped and crinkled with the terminal bud deterioration.
- Buds and blossoms fall prematurely in some crops.

Magnesium

- Because Mg is a mobile element and part of the chlorophyll molecule, the deficiency symptom of interveinal chlorosis first appears in older leaves. Leaf tissue between the veins may be yellowish, bronze, or reddish, while the leaf veins remain green. Crops, such as potatoes, tomatoes and cabbage show orange-yellow colour with green veins.
- In severe cases, symptoms may appear on younger leaves and cause premature leaf drop.
- Symptoms occur most frequently in acid soils and soils receiving high amounts of K or Ca.

Sulphur

- Younger leaves are chlorotic with evenly lightly coloured veins. In some plants, the older leaves may show symptoms first. However, deficiency is not commonly found in most plants.
- Growth rate is retarded and maturity is delayed.
- Plant stems are stiff, thin and woody.
- Symptoms may be similar to N deficiency and are most often found in sandy soils that are low in organic matter and receive moderate to heavy rainfall.

Zinc

- Interveinal chlorosis occurs on younger leaves, similar to Fe deficiency. However, Zn deficiency is more defined, appearing as banding at the basal part of the leaf; whereas Fe deficiency results in interveinal chlorosis along the entire length of the leaf.
- In vegetable crops, colour change appears in the younger leaves first. The new leaves are usually abnormally small, mottled and chlorotic.
- In citrus, irregular interveinal chlorosis occurs with small, pointed, mottled leaves. Fruit formation is significantly reduced.
- In legumes, stunted growth with interveinal chlorosis appears on the older or lower leaves.

Copper

- Reduced growth, distortion of the younger leaves and possible necrosis of the apical meristem.
- In trees, multiple sprouts occur at growing points, resulting in a bushy appearance. Young leaves become bleached and eventually, there is defoliation and dieback of twigs.

Iron

- The most obvious symptom of iron deficiency in plants is commonly called leaf chlorosis. This is where the leaves of the plant turn yellow, but the veins of the leaves stay green.

- Typically, leaf chlorosis will start at the tips of new growth in the plant and will progress to older leaves on the plant as the deficiency gets worse.
- Other signs can include poor growth and leaf loss, but these symptoms will always be coupled with the leaf chlorosis.

Manganese

- Symptoms first appear as chlorosis in young tissues. Unlike Fe chlorosis symptoms in dicots, Mn chlorosis shows up as tiny yellow spots.
- In monocots, greenish-grey specks appear at the lower base of younger leaves. The specks may eventually become yellowish to yellow-orange.
- In legumes, necrotic areas develop on the cotyledons, a symptom known as marsh spots.

Boron

- Generally, B deficiency causes stunted growth, first showing symptoms on the growing point and younger leaves. The leaves tend to be thickened and may curl and become brittle.
- In many crops, the symptoms are well-defined and crop-specific, such as:
- Crooked and cracked stem in celery
- Black hearts in beets
- Distorted and lumpy fruit in papaya
- Splitting of calyx in carnation
- Pith in hollow stem in cabbage, broccoli and cauliflower

Molybdenum

- Deficiency symptoms resemble those of N because the function of Mo is found to be essential to assimilate N in the plant. Older and middle leaves become chlorotic and the leaf margins roll inwards.
- In contrast to N deficiency, necrotic spots appear at the leaf margins because of nitrate accumulation.

- Deficient plants are stunted and flower formation may be restricted.
- Mo deficiency can be common in nitrogen-fixing legumes.

Chlorine

- Chlorosis of younger leaves and wilting of the plant.
- Deficiency seldom occurs because chlorine is found in the atmosphere and rainwater.

Nickel

- Nickel deficiency will display visual symptoms typically in the old leaves of the plants as nickel is a mobile element.
- In woody ornamentals, symptoms occur in newly emerged shoot and may include shortened internodes (giving a rosetting appearance to the plant), weak shoot growth, death of terminal buds and eventual death of shoots and branches.
- The leaves develop a condition called "mouse-ear" in which the leaflets are small with rounded tips.

d) Nutrient Toxicities / Excesses

Damage from excessive nutrient levels can occur in crops and ornamental plants as the result of over-application of fertilizers or manures. Nitrogen toxicity is most typical under hot, dry conditions and plants turn an overly-deep shade of green. Lesions often occur on the stems of annual seedlings and these can be confused with canker diseases. Similarly, twisting and distorting of mature tomato plants that experience ammonium toxicity may appear similar to symptoms caused by viruses. Ammonium toxicity can be a problem in greenhouse soils because of the lack of specific microorganisms that convert ammonium to nitrite and then to nitrate. Micronutrient toxicities are common in many production systems. Symptoms often include chlorosis or necrosis on leaf margins or tips, but leaf spotting, flecking and other symptoms can occur. As noted above, many nutrient toxicities are triggered by excessively low or high soil pH. Micronutrient toxicities are particularly common in greenhouse flower crops. For example, Fe and Mn toxicity often occurs in greenhouse crops, when the growing medium has a low pH. Excessive micronutrients can also occur, when irrigation water or soil has significant concentrations of micronutrients. Other elements that are not plant nutrients, e.g., lead and arsenic and other heavy metals can also be toxic to plants.

Toxicity symptoms

Nitrogen

Plants usually dark green in colour, abundant foliage but usually with a restricted root system. Flowering and seed production can be retarded.

Phosphorus

No primary symptoms yet noted. Sometimes copper and zinc deficiency occurs in the presence of excess phosphorus.

Potassium

Excess potassium will not produce any toxicity. Excess potassium may lead to magnesium deficiency and possible manganese, zinc or iron deficiency.

Zinc

Excess zinc commonly produces iron chlorosis in plants.

Iron

Iron toxicity is related to the plant's excessive iron absorption. This excessive absorption damages the plant cells. Symptoms appear initially on younger leaves, where the element concentrates in small brown dots. This phenomenon is known as "bronzing". Under extreme situation, the leaves may become chlorotic and at advanced toxicity stages, necrosis may occur, i.e. the leaves dry and eventually die.

Boron

Yellowing of leaf tip, followed by progressive necrosis of the leaf beginning at tip or margins and proceeding towards midrib.

Sulphur

Reduction in growth and leaf size. Leaf symptoms often absent or poorly defined. Sometimes interveinal yellowing or leaf burning.

Copper

Reduced growth followed by symptoms of iron chlorosis, stunting, reduced branching, thickening and abnormal darkening of rootlets.

Calcium

No consistent visible symptoms. Usually associated with excess carbonate.

Molybdenum

Rarely observed. Leaves turn golden yellow.

Magnesium

Very little information available on visual symptoms.

Manganese

Sometimes chlorosis or uneven chlorophyll distribution. Reduction in growth.

Chlorine

Burning or firing of leaf tip or margins. Bronzing, yellowing and leaf abscission and sometimes chlorosis. Reduced leaf size and lower growth rate.

e) Fruit Cracking

Fruit cracking or splitting is a problem that may be associated with a number of environmental, cultural, genetic and anatomical factors. Cracking occurs when there is a rapid influx of water and solutes into the fruit at the same time that other factors interact to reduce the strength and elasticity of the fruit skin. Factors thought to promote this disorder include irregular watering, high temperatures and high irradiance, excessive growth rates, high difference between day and night temperature, high humidity and poor calcium nutrition. Problems with fruit cracking are typically associated with greenhouse environments, but the causal factors suggest that injury of susceptible plants might readily occur in growth chambers. Methods proposed to prevent cracking include selection of non-susceptible cultivars, consistency in watering, good calcium nutrition and possibly the use of calcium or gibberellic acid.

f) Salt Injury

Symptoms of "salt" or "fertilizer" injury may be variable but usually include: slow and spotty seed germination, sudden wilting, stunted growth, marginal burn on leaves (especially lower, older leaves), leaf yellowing, leaf fall, dead roots, restricted root development and sudden or gradual death of plants. The tops of plants may show "salt injury" while the roots are apparently unaffected. In this case, the soluble salts enter the roots and are moved through the plant vessels to the leaves where the water evaporates and gradually concentrate the salts to toxic levels. When soil soluble salts are excessively high, the roots are unable to absorb water and the plant wilts.

g) Biuret injury

Biuret is formed, when urea is subjected to heating. More biuret is formed, when urea is stored in moist places. The biuret content should be less than 0.5 per cent. When urea is applied as foliar spray with more than 1.0 per cent biuret content, it produces injury in plants. The symptoms of injury includes, yellowing of leaves and cupped growth in citrus, coffee and pine apple. Sometimes, interveinal chlorosis, stunted growth and unfolded leaves also observed in certain crops due to biuret injury.

h) Transplantation Shock

Transplantation shock, is a term that refers to a number of stresses occurring in recently transplanted crops, trees and shrubs. It involves failure of the plant and consequently, the plant becomes poorly established. New transplants do not have extensive root systems and they are frequently stressed by lack of sufficient water. Plants suffering from water stress may be more susceptible to injury from other causes such as the weather, insects, or disease.

Leaf scorch is a common symptom of transplantation shock. Leaf scorch first appears as a yellowing or bronzing of tissue between the veins or along the margins of leaves of deciduous plants. Later, the discoloured tissue dries out and turns brown. Other symptoms of transplant shock appear as wilting of leaves (especially on recent transplants), yellowing and leaf rolling or curling.

Overall plant growth is greatly reduced which is reflected in the length of the new growth. Poorly growing plants have extremely shortened internodes, resulting in shortened branch tips relative to those of an unstressed plant. New leaves or needles of a stressed plant are smaller than normal.

i) High Irradiance

Sudden or large increases in irradiance can often damage leaf tissues. Such a situation can occur, when plants are transferred from low light levels in green houses during the winter to a growth chamber with high light levels. Typically, these changes cause a bleaching or bronzing of the exposed leaves, but the damage is generally transient and new developing leaves seem to adapt to the higher light levels.

j) Vitrification

Vitrification, also called glassiness or water soaking, is a physiological disorder that almost exclusively affects plants maintained in sterile culture. Vitrification has been observed in a number of species and most commonly is characterized by a translucent or water-soaked appearance of leaves and stems. In addition, leaves can become enlarged and malformed, exhibiting wrinkling or curling. Because of their extreme turgidity, the translucent leaves become very brittle and can break easily. On a cellular level, vitrified leaves lack palisade tissue (having only a spongy mesophyll) and generally exhibit defective lignifications of vessels and tracheids. Though the mechanism responsible for the disorder is not known, one hypothesis is that vitrification results from a burst of ethylene triggered by the peroxidase -IAA-oxidase system.

k) Leaf Pleating

Leaf pleating is characterized by a crinkled like malformation on the leaves of monocots. This disorder has been observed on several sympodial orchid species and ornamental grasses. The disorder seems to occur, when leaf tips fail to emerge freely from the leaf sheath and become compressed within the sheath. When the leaf finally breaks out of the sheath, the pleated pattern characteristic of this disorder can be observed. Although the cause of this disorder is not known, it has been attributed to water stress of new developing growth.

l) Deficiencies in Available water

Drought stress occurs, when plant roots are not absorbing enough water for their needs. There are many causes for drought stress.

Drought may be of two kinds: short-term and long-term. An example of a short-term drought is the length of a growing season. A long-term drought lasts more than one growing season. While a short-term drought can damage the plants, the long-term droughts are more harmful due to the chronic moisture stress. Symptoms are the plant's reaction to stress and provide clues during diagnosis. Following are some common symptoms of drought stress.

Symptoms on whole plant

- The pattern of plant damage or death occurs from the top of the plant.
- Plants wilt. One of the first symptoms of drought-stressed plants is the loss of turgidity. Plants or plant parts become limp and droopy.
- Plants show a decrease in growth or have no growth, both in girth and in length. A way to verify this on woody plants is to check the length of the growth increments, the amount of growth produced in each season. Beginning at the tip of a twig, move along the twig toward the trunk. Look for the first set of "wrinkles". The distance from the tip to the first set of wrinkles shows the amount of growth produced during the most recent growing season. Look for the next set of wrinkles. This shows the amount of growth produced by the plant during the previous season.
- Plants appear chlorotic.
- Tree canopy may be thin.
- Plants may leaf out, then die later in the growing season as a result of depleted food reserves. This may occur during or even a few years after a drought event.
- Gummy exudates appear on twigs, branches and trunks.
- Suckers develop on branches and trunk.
- Wood or bark cracks.
- Stems and twigs die with the outermost and upper ones dying first.
- Entire plants may die, as the result of root death from dehydration.

Symptoms on leaves

- Leaves are smaller than normal.
- Deciduous leaves turn brown from the outside edges inward and also in between the veins ("scorch"). This symptom occurs because these areas naturally have the least amount of moisture in the leaf.
- Evergreen leaves turn yellow, red or red-purple.
- Leaves roll up and/or are misshapen.
- Leaves drop prematurely. They may or may not turn colour prematurely before dropping.
- Leaves remain attached to tree, even though brown.
- Leaves are dull in appearance rather than shiny.
- Leaves may turn blue-green.

Flower and Fruit Symptoms

- Flowers fail to open properly.
- The flowering period is shorter than normal.
- Fruit drops early.
- Fruit and seed production may be reduced or absent.

m) Excessive Quantities of Water

Excessive soil moisture can result in reduced oxygen availability to roots. Oxygen is a principal component for the physiological uptake of water into the roots. Therefore, similar to drought, a primary symptom of flooding is wilt. During acute water excess (flooding), roots are subjected to low oxygen status and cell weakening and/or death occurs. Symptoms include discolored and/or water-soaked and mushy roots. Under these conditions, roots can become weakened and predisposed to invasion by pathogens such as *Phytophthora* spp. Flooding events also predispose plants to root diseases caused by certain pathogens that have flagella to swim towards host roots.

In the case of excess of water, plants appear stunted and have underdeveloped shoots. In severe cases, bleeding cankers on stems can occur. Adventitious roots may form at the root crown. Bark can split and wood may become water-soaked and discoloured. Edema or corky, blister-like swelling can occur on the underside of leaves on plants growing in waterlogged soils. Edema can be worse during cloudy, overcast periods. In areas where waterlogged soils prevail for long periods of time, an odour of rotten eggs may be noticeable, due to sulphur gas production in the anaerobic soil.

n) Wind Injury

The air moving horizontally on the surface of the earth is known as wind. Vertically or nearly vertical movements of air resulting from convection, turbulence or any other cause is known as air current. Wind increases the transpiration and intake of CO_2. The turbulence created by wind increase CO_2 supply and increase in photosynthesis. When wind is hot, desiccation of the plants takes place, because humid air in the intercellular places is replaced by dry air. The hot and dry wind makes the cells expanding and early maturity, resulting in the dwarfing of plants. Under the influence of strong wind, the shoots are pressurized and get deformed. Strong winds produce loading of crops. The coastal areas affected by strong wind bring salt and make the soil unsuitable for growing plants. Strong winds affect the plants life both mechanically and physiologically.

o) High Temperature Injury

Some plant species can be very sensitive to high temperatures. In plants adapted for cooler climates, physiological changes can result in response to excessively high temperatures. For example, shoots and/or roots may stop growing if high temperatures prevail for an extended period of time. Roots may die. If high temperatures are coupled with low soil moisture, plants may exhibit scorching on the margins of the leaves, premature leaf drop and in severe cases, entire plant death. Sometimes physiological changes result in abnormal colour or growth habits. Another common occurrence is the effect of high temperatures on pollination. Many food crops species are highly subject to poor pollination during periods of high temperatures. Heat injury symptoms are as follows :

- Retarded growth and undersize or failure of fruit to mature;
- Localized killing of tissues or sunburn or sun scald of leaves, flowers or fruits;
- Localized killing of stem tissues or formation of heat cankers;

- Defoliation or premature shedding of leaves;
- Premature ripening of fruits; and
- Death of the plant as the result of a general heat necrosis.

p) Low Temperature Injury

Low temperature impacts on plants can be classified as chilling or freezing. Chilling injury or chilling stress, as it is often termed, is associated with temperatures above freezing but low enough to cause injury. Chilling injury is due to a sudden drop in temperature during an active period of plant growth or development. The greater the drop in temperature from normal, the more likely injury will occur. The extent of the injury may cause a reduction in photosynthesis or other metabolic processes, an alteration in growth or death of the exposed plant tissue. For example, a substantial drop in temperature during new shoot growth and leaf expansion may result in damage to the terminal meristem and a loss in shoot growth. Other signs of chilling injury may include wilting, desiccation and/or a physical distortion of plant parts. Chilling injury is less serious than freezing injury on most of the herbaceous and woody plants.

Freezing injury is caused by subfreezing temperatures. The impact of subfreezing temperatures may be felt from as little as one overnight episode or from a more sustained exposure episode. Plants subjected to subfreezing temperatures may exhibit ice formation in the vascular vessels, in the spaces between the cells and/or within the cells. The extent and pattern of ice formation vary with the plant tissue and its condition. Ice formation in the spaces between cells is usually not fatal, but it causes water to leave the cells, causing dehydration. If the freezing episode is cold and long enough, it can result in freezing within the cells. Internal or intracellular freezing ruptures membranes and leads to cell death. Freezing damage begins with dehydration and ultimately ends in death of plant. Typical symptoms of freezing injury are a blackened/ brownish discolouration or bleaching of plant tissue. The severity of the freezing damage depends on the type of frost injury and the plant parts affected. If the freezing injury kills a significant number of buds or cambial tissue, the plant may die outright or suffer so much crown die-back that it becomes unusable or unacceptable. More commonly, however, freezing injury may be limited to flower buds and minor shoot dieback. This damage may require corrective pruning and time to allow the plant to grow out of the damage.

q) Oedema

Oedema, is a physiological disorder that develops when a plant absorbs water faster than it can be lost normally from leaf surfaces. The term itself means 'swelling', which describes its initial stages. Excess moisture builds up in the plant and causes swellings that appear initially as pale-green or water-soaked blisters or bumps, primarily on the undersides of leaves.

Oedema appears as small, sometimes corky blisters which form on the lower surface of leaves or needles. These blisters may eventually harden to form white, tan, or brown war like corky areas on the lower leaf surface. In severely affected plants, these corky growths also form on petals, petioles and stems. As injury continues, leaves turn yellow, droop, and fall off. Plants become spindly and growth ceases.

Oedema is most prevalent in the late winter especially during extended periods of cool, cloudy weather. It is likely to develop when the soil is warm and moist and the air is cool and moist. This environment results in rapid water absorption from the soil and slow water loss from the leaves. These conditions are most frequently encountered in greenhouses or indoor situations rather than outdoors. Geraniums, rhododendrons, begonias and certain fleshy leaved plants are particularly sensitive to conditions which lead to the development of oedema, although almost any broad leaved plant may be affected. Vegetables, such as cabbage and tomato can also be affected.

Overwatering, high humidity and low light intensities are factors which favour the development of oedema. Avoid overwatering susceptible plants especially during the winter months when they should be kept slightly on the dry side. Keep the relative humidity below 70% in the winter. Improve the flow of air over the leaves by spacing plants farther apart and increasing ventilation. Affected plants often recover from oedema with the return of more favorable growing conditions in spring and early summer. Gradually increasing the exposure to sunlight is beneficial and will help avoid sun-scorch, when plants are moved outdoors.

r) Lightning Injury

The most severe damage to plants by lightning may be caused by the extreme heat and shock waves generated by the current, although other damaging effects probably occur. The return stroke propagates up the leader channel at a velocity of about 1/10 the speed of light, making the trip from the ground through or around the plant and back to the cloud in about 100 millionths of a second. The energy in this current produces temperatures in the channel greater than 50,000 degrees Fahrenheit in millionths of a second. The heat turns plant fluids into steam and burns plant cells and tissues, leading to a wilting symptom and blackened, necrotic, scorched tissues, including roots, stems, branches and fruits.

The tremendous heat energy is produced in millionths of a second, which is far too quick for the air to expand gradually. The pressure produced along the channel by this intense heat is on the order of 10 to 100 times normal sea-level pressure. This high-pressure region rapidly expands outward, causing compression of the surrounding air. This region of compression propagates outward as a shock wave. The pressure can cause the pith of a plant stem to explode out through holes in the stem, it can cause immediate drop of all fruits (ripe or not) and it can split a tree trunk.

Symptoms

- There is rapid wilting or collapse of plants or stems, in combination with structural damage or carbonization (burning or black scorching) of the internal stem tissues and/or browning or blackening of leaves, fruits, or stems.
- Symptoms develop rapidly; they appear in days rather than weeks.
- Symptom onset coinciding with recent thunder storms.
- A circular area of plants in a field or orchard are affected.
- There is premature mass dropping of green fruit.
- There are burns or strange scars on plants or organs.
- Roots may be blackened or cooked.
- Interior plant tissues, such as stem pith, xylem and phloem are blown out of the stem.
- In crops, such as sugar beets, potatoes or sweet potatoes, underground tubers or roots are cooked.

s) Light Deficiency

As the light intensity and duration level drops from the minimum to zero or complete darkness, the plant may undergo gradual, formative or structural changes, including alteration of colour and peculiarities of structure which are characteristic of the condition known as etiolation.

The changes which are characteristic of etiolation are as follows:

- The abnormal elongation of stems (internodes) and petioles or of leaves that are normally elongated and have a basal or intercalary meristem, thus producing shoots that may be characterized as "spindling." This modification occurs in plants with stems that are normally elongated and also in those of the rosette habit.
- The pronounced reduction in the size of the leaves (shade leaves in many cases are larger than sun leaves). This feature is characteristic of dicotyledons, while monocotyledons generally show an elongation and narrowing of the leaves, although deviations from this behaviour may be found in both groups.
- The reduction in the amount of chlorophyll or its complete disappearance (in darkness) and consequently the slowing up of photosynthesis.
- The suppression of the reproductive function, as illustrated by sterility, lessened flower formation or the complete absence of blossoms.
- The stems are more slender and leaves thinner, cell walls are more delicate and there is a poorer development of mechanical tissue.
- Plants grown under poor light conditions wilt more readily than normal plants when exposed to bright light.

t) Chemicals

(i) Herbicides

Herbicides, when used properly, rarely cause problems on non target plants. However, these products can cause injury when applied inappropriately, when they turn into a gas, or when they are blown by the wind away from the targeted area. Most cases of herbicide injury in field crops can be traced to six herbicide classes or families:

triazines, ALS-inhibitors, HPPD-inhibitors, plant growth regulators, PPO-inhibitors and glyphosate. Some common symptoms associated with each of the herbicide groups are given below:

- **Triazines** : Triazines are photosynthesis-inhibiting herbicides that block the photosynthetic process, so captured light cannot be used to produce sugars. The plant slowly starves to death due to lack of energy. In broad leaved plants, early seedling growth appears normal, but shortly after emergence (when energy reserves in cotyledons are depleted), leaves become mottled, turn yellow to brown and die. In most cases, the oldest leaves turn yellow on the leaf margins first, the veins remain green and eventually the plant turns brown and dies. Herbicides, such as atrazine and simazine could cause this kind of injury.
- **ALS-inhibitors** : These herbicides work by interfering with one or more key enzymes that catalyze the production of specific amino acids in the plant. When a key amino acid is not produced, the plant's metabolic processes begin to shut down eventually causing plant death. Plants that are sensitive to these herbicides stop growth almost immediately after foliar treatment, seedlings die in three to seven days, established perennials due in two to four weeks. Symptoms include: stunted, yellow, purple veins, dead growing point, roots malformed (bottle-brush). These herbicides have systemic activity throughout the plant and young leaves are affected first. Grass plants may be stunted with interveinal yellowing (chlorosis) or purpling.
- **HPPD-inhibitors** : These products are referred to as "bleachers" since they interfere with normal chlorophyll formation. Symptoms are very evident and easy to identify. Affected plants turn white or show bleached leaves and eventually die if concentration of herbicide is high.
- **Plant growth regulators** : In most cases, plant growth regulators damage occurs not from carryover but from herbicides that are sprayed in adjacent fields and then drifts onto the susceptible crop. Plant growth regulators upset normal growth as follows: cells of leaf veins rapidly divide and elongate, while cells between veins cease to divide. This results in long, narrow, strap-like young leaves. Water content increases, making treated plants brittle and easily broken. Cell division and respiration rates increase and photosynthesis decreases. Food supply of treated plants is nearly

exhausted at their death. Roots of treated plants lose their ability to take up soil nutrients and stem tissues fail to move food effectively through the plant. The killing action of growth-regulating chemicals is not caused by any single factor, but results from the effects of multiple disturbances in the treated plant. Symptoms include: broadleaf plant leaves become cupped, crinkled, puckered, strap-shaped, stunted and malformed; leaf veins appear parallel rather than netted and stems become bent, twisted and brittle with shortened internodes.

- **PPO-inhibitors** : These herbicides are referred to as contact herbicides and they kill weeds by destroying cell membranes. They appear to burn plant tissues within hours or days of application. Good coverage of the plant tissue and bright sunlight are necessary for maximum activity. The activity of these herbicides is delayed in the absence of light. All contact herbicides cause cellular breakdown by destroying cell membranes, allowing cell sap to leak out. Affected plants initially have a "watersoaked" appearance, followed by rapid wilting and "burning," or leaf speckling and browning. Plant death occurs within a few days.

Glyphosate

Plant foliage, especially new growth, will first become yellow, then turn brown and die within a week or so after herbicide application. Sometimes new leaves on sensitive plants exhibit a bright yellow or even white appearance which can be confused with injury from other herbicide groups.

(ii) Pesticides

Insecticides, fungicides and other chemicals used in the control of pests can cause problems on both targeted and non-targeted plants. Several factors can influence whether the pesticide application is the solution or becomes part of the problem. As with any chemical application, injury will depend on the type, concentration and application rate of the compound, the plant species, its stage of growth and physiological condition and atmospheric conditions–light, temperature, humidity and wind. Finding the perfect time to apply pesticides is always a challenge. Proper handling of the pesticides and close monitoring of weather conditions can avoid application related injuries. Application injuries exhibit the same symptoms as nutrient disorders and other chemical applications. In most instances, leaf chlorosis, marginal and/or spotted necrosis and total leaf necrosis are the visible symptoms.

u) Air Pollution Injury

Air pollution injury on plants has been carefully documented in recent years. In some areas, injury on certain species has been serious and has represented a limiting production factor. This type of damage is more noticeable, but probably not as great as the sub-lethal chronic injury that occurs over much larger areas. Chronic damage is more difficult to diagnose and evaluate. Diagnosticians should be aware that pollutants are mixed in the atmosphere and that these mixtures may produce symptoms that vary from those caused by a single compound under laboratory conditions.

The effects of pollution on plants include mottled foliage, "burning" at leaf tips or margins, twig dieback, stunted growth, premature leaf drop, delayed maturity, abortion or early drop of blossoms and reduced yield or quality. In general, the visible injury to plants is of three types: (i) collapse of leaf tissue with the development of necrotic patterns, (ii) yellowing or other colour changes and (iii) alterations in growth or premature loss of foliage. Injury from air pollution can be confused with the symptoms caused by fungi, bacteria, viruses, nematodes, insects, nutritional deficiencies and toxicities and the adverse effects of temperature, wind and water.

v) Bud Sports or Bud Mutations

When mutations occur and suddenly appear as a chance in the branch of a plant, it is called **bud sport or bud mutations**, because they appear to have originated within a single bud.

- Detection of a new mutant within a clone may require a series of vegetative propagated generations and multiple propagations from many buds of the same plant.
- Many 'sports' have become commercial cultivars. Mutations may affect fruit (colour, shape, time of maturity), tree structure (spur type), time of bloom and as a host of other traits. Sometimes these mutants have highly useful horticultural traits and have given rise to important new cultivars.

- On the other hand, mutations may be undesirable and give rise to misshapen fruits, low production and susceptibility to diseases.

w) Chimeras

Chimeras are botanical abnormalities that are often confused with nutritional or chemical disorders. "Chimera" is a term used to describe a single plant with two genetically different tissue types. Leaf variegation is the most common example of chimeras in plant species. The difference in foliage colour and the banding of those colours are due to cell mutations in the meristematic tissue layers. Chimeras can be stable, in that the genetic differences are consistent and reproducible. These chimeras have spawned numerous of our variegated plant cultivars. Chimeras can also be unstable and unpredictable, surfacing sporadically on either shoots or individual leaves.

x) Allelopathy

Allelopathy refers to the beneficial or harmful effects of one plant on another plant, both crop and weed species, from the release of biochemicals, known as allelochemicals, from plant parts by leaching, root exudation, volatilization, residue decomposition and other processes in both natural and agricultural systems. Allelochemicals are a subset of secondary metabolites not required for metabolism (growth and development) of the allelopathic organism. Allelochemicals with negative allelopathic effects are an important part of plant defense against herbivory (i.e., animals eating plants as their primary food).

Allelopathic inhibition is complex and can involve the interaction of different classes of chemicals, such as phenolic compounds, flavonoids, terpenoids, alkaloids, steroids, carbohydrates and amino acids with mixtures of different compounds sometimes having a greater allelopathic effect than individual compounds alone. Furthermore, physiological and environmental stresses, pests and diseases, solar radiation, herbicides and less than optimal nutrient, moisture and temperature levels can also affect allelopathic weed suppression. Different plant parts, including flowers, leaves, leaf litter and leaf mulch, stems, bark, roots, soil and soil leachates and their derived compounds, can have allelopathic activity that varies over a growing season. Allelopathic chemicals or allelochemicals can also persist in soil, affecting both neighbouring plants as well as those planted in succession. Although derived from plants, allelochemicals may be more biodegradable than traditional herbicides, but allelochemicals may also have undesirable effects on non-target species, necessitating ecological studies before widespread use.

y) Acid Rain Injury in Plants

Acid rain forms when sulphur dioxide and nitrogen oxide reacts with chemicals like water, oxygen and carbon dioxide in the atmosphere to form sulphuric acid and nitric acid. Water containing these acidic compounds falls back to the earth as rain, harming plants and other immobile objects below. Although the acid from acid rain is weak, normally no more acidic than vinegar, it can seriously alter the environment, damaging plants and aquatic ecosystems. Dead or dying trees are a common sight in areas affected by acid rain. Acid rain leaches aluminium from the soil. Aluminium may be harmful to plants as well as animals. Acid rain also removes minerals and nutrients from the soil that trees need to grow. At high elevations, acidic fog and clouds might strip nutrients from trees' foliage, leaving them with brown or dead leaves and needles. The trees are then less able to absorb sunlight, which makes them weak and less able to withstand freezing temperatures.

z) Animal Damage

The issue of food supply and demand is a concern for animal damage. When forage becomes limited, many of our ornamentals become prime targets for deer, rabbits, squirrels, mice and other rodents. Squirrels may feed on the petioles of newly expanded leaves. Rabbits, mice and other rodents focus on the ground-level portions of deciduous trees and shrubs. Habitat reduction is the first line of defense against small rodents. Unfortunately, many of our prized plant compositions also make for ideal cover. Minimizing cover may deter feeding but does not eliminate it. Caging plant crowns, wrapping wire mesh around the bases of the plants, applying feeding repellents and using poison baits work reasonably well.

❑❑❑

II. Fruits

1

Mango

a) Spongy Tissue

It is a major problem in Alphonso. It is a physiological disorder in which fruit pulp remains unripe because of unhydrolyzed starch due to physiological and biochemical disturbances.

Management

Use of mulching and post harvest exposure to lower the temperatures between 10-15°C for 10-18 hours has been useful in reducing the malady.

b) Fruit Drop

In mango, there is a heavy drop of hermaphrodite flowers and young fruits amounting to 99% or more. The commercially grown varieties, Langra and Deshehari are more susceptible to drop in the first three weeks of April. The fruit drop is more or less a continuous process and can be classified into three phases, viz. (i) Pinhead drop, (ii) Post-setting drop and (iii) May-month drop. The fruit drop in first two phases are insignificant compared to the third phase which affects the final yield significantly.

Management

Regular irrigation during fruit setting and development period can reduce this problem. Application of plant growth regulators like NAA @ 40 ppm or 2, 4 D @ 20 ppm at 6 weeks after fruit set, (at pea stage) reduce fruit drop considerably. In addition to the above, finely and effective control measures against major pest and diseases also reduces the fruit drop.

c) Black Tip

Black tip is a serious disorder, Coal fumes of brick kilns containing sulphur dioxide, ethylene and carbon monoxide are observed to be responsible for black tip. The damage has been noticed in the mango orchards located up to 200 metres of distance from brick kiln. The damage to the fruit gets initiated right at marble stage with a characteristic yellowing of tissues at distal end. Gradually, the colour intensifies into brown and finally black. At this stage, further growth and development of the fruit is retarded and black ring at the tip extends towards the upper part of the fruit.

Management

Planting of mango orchards in North-South direction and 5-6 km away from the brick kilns may reduce incidence of black tip to a greater extent. The incidence of black tip can also be minimized by spraying Boric acid (0.1%) or 2% sodium carbonate. The first spray of Borax should be done positively at pea stage followed by two more sprays at 15 days interval.

d) Mango Malformation

Mango malformation is of two types viz. vegetative and floral. The vegetative malformation generally affects seedlings of young plants in which there is a swelling of buds and formation of small shoots with short inter nodes at the apical portion and given an appearance of witches broom like structure. In floral malformation, panicles become deformed, axis become short and rachis thick, due to this, inflorescence may look like a cluster. Malformed panicles have bigger flowers than the normal flowers and are mostly male.

Management

Spraying of Planofix (200 ppm) during the first week of October followed by deblossoming at bud burst stage is recommended as a remedial measure against malformation.

e) Clustering in Mango ('Jhumka')

A fruiting disorder, locally known as 'Jhumka', is characterised by the development of fruitlets in clusters at the tip of panicles. Such fruits cease to grow beyond pea or marble stage and drop down after a month of fruit set. Absence of sufficient population of pollinators in the orchards is the major reason. The other reasons causing the disorder are old and overcrowding of trees, indiscriminate spraying against pests and diseases, use of synthetic pyrethroids, monoculture of Dashehari and bad weather during flowering.

Management

Introduction of beehives in the orchards during flowering season for increasing the number of pollinators and restrict insecticidal sprays at full bloom to avoid killing of pollinators.

f) Irregular Bearing in Mango

The mango industry as a whole is threatened by a physiological disorder called irregular bearing or biennial or alternate bearing. The tendency of mango tree to bear heavily in one year and low or no bearing during next one or two years is called as alternate bearing in mango; when mango bears heavily, it is called 'on' year and when mango bears less or no crop, it is called 'off' year.

Generally, it occurs after the age of 10 years. Upto this stage, the plants put forth new shoots as well as the flowers simultaneously. The new shoots are used in the next year for producing flowers. This is due to the ability of the older trees to produce sufficient amount of leafy shoots in addition to biomass.

External factors

i) Climate

- Prevailing of high temperature during fruit set causes heavy drop of small fruits.
- Low temperature at the time of flowering suppresses the flowering.
- Rainfall at the time of flowering, wash out sigmatic fluid and affect the process of pollination.

ii) Pest and Diseases

Diseases like powdery and pest like mango hopper cause flower and fruit drop.

iii) Nutrition

Less water availability to the plants in "on" year produces sparse vegetative growth and as there is no vegetative growth, the next year will remain off year.

Internal factors

i) Varietal character

Genetic makeup of a plant decides its characters. Regular or alternate bearing is also specific character. Some varieties are regular bearing and some other bear in alternate years.

Alternate bearer - Alphonso

Regular bearer - Neelum, Banglora

ii) Auxin content

Hormonal imbalance at the time of fruit bud differentiation, flower production, fruit set, results into "off" year.

iii) C: N ratio

If C: N ratio is 60:40, then there will be flowering and if it is 40:60, there will be vegetative growth.

iv) Sex differentiation

There are around 1000 panicles containing individually 30-3000 flowers. Out of these, very few flowers are perfect which give rise to fruits. Large number of flowers do not set fruits due to structural defects. Perfect flowers: Alphonso-10%, Langra-66%, Neelum-16%

v) Age of plant

Mango trees, when young i.e., upto 10 years, produce fruits regularly. After 10 years, it starts bearing fruits in alternate years.

vi) Ringing

It is done during off year in the month of August in central part. By ringing, the carbohydrate flow is obstructed on the upper side of the ring which results in producing more flowering shoots.

vii) Deblossoming

If half of the total number of flower clusters are removed from the tree in "on" year as soon as they emerge, the food reserve of these deblossomed shoots will be utilized by the tree, producing vegetative growth in summer which ultimately results in producing flower clusters in "off" year.

viii) Smudging

Smudging is nothing but treating the orchard with smoke. The smoke is produced by half burning material like grass etc. In Phillipines, smudging during night for 8 days induces flowering in November-December.

ix) Use of Cultar

Paclobutrazol, an anti-gibberellin has profound effect on increasing the fruit set, besides giving better fruit size and quality. The rate of Paclobutrazol application depends on the size of tree canopy as well as on mango cultivars. For most cultivars, the rate of Paclobutrazol applied is generally determined by multiplying the diameter of tree canopy (expressed in meter) with 1.0-1.5 g of active ingredients of paclobutrazol. Paclobutrazol (Cultar) should be used in bearing trees and the dose varies from 1ml to 10ml based on the canopy spread. Normally, Paclobutrazol is applied 90 days ahead of flower opening. Required quantity of Paclobutrazol is dissolved in 10-15 litres of water and drenched in holes at 8-10 places in circle half way of canopy spread.

x) Spraying of chemicals

The foliar spraying of potassium nitrate at 4 per cent on new leaves with full green colour and buds at 14 days interval is effective in induction of off-season cropping, ensuring annually two mango harvesting seasons. Three times foliar spraying of KH_2PO_4 (1%) + KNO_3 (1%) during second week of September, October and November induces earlier flowering and fruiting in mango Alphonso.

g) Fruit Pitting in Mango

Initially, small sunken pits on peel of fruits develop which attain half grown size. Pits appear in almost equal proportion on all sides of fruits and increase in size as fruits grow and reach the maturity. Development of pits on fruit peel gives nude and ugly look. Almost all the mango varieties are affected by it, but severity is comparatively higher in Dashehari, Amrapali and Chausa. Further, thick-peeled varieties are more prone to this disorder than thin-peeled ones. Preliminary observation indicates that these pits also sometimes serve as foci centre for attack of fruit fly. After which, exudation of gum is quite common from the affected fruits. The gum trickles down the affected fruits in the form of flakes. In severe infestation, fruits affected by pitting and fruit fly usually drop down and get rotten. Rotten fruits show foul smell. However, those affected alone by 'pitting', neither drop down nor get rotten. However, such fruits affect the consumers' acceptability. It may be due to:

- Poorly managed orchards
- Moisture stress during fruit development
- Deficiency of boron in soil
- Hormonal imbalance

Management

Avoid moisture stress during fruit development. Application of 300-400 g borax / tree to fully bearing trees helps reduce the incidence of fruit pitting to a greater extent. In 6-10 year old plants, borax @ 150-200 g/tree is sufficient to reduce its incidence. Most appropriate time of application of borax is mid-April to April-end, i.e. immediately after fruit setting or when fruits are at marble-sized stage.

Borax should be applied in the form of ring around the tree trunk. After thoroughly mixing borax with soil, a light irrigation is necessary. However, foliar application of borax is not as effective as soil application. Application of borax also checks internal brown necrosis or brown rot and black tip in mango.

Irrigation management and maintaining stable moisture level during fruit growth and development will also reduce the fruit splitting.

h) Sap-injury in Mango

Sap-injury is an important post-harvest problem in mango. The mango fruit has a network of branching fruit ducts, which are present in both fruit and stalk and penetrate the transition zone between the fruit and its stalk. Several large fruit ducts continue into the stalk but end a short distance beneath the abscission zone. These fruit ducts contain a viscous, caustic liquid referred to as mango sap. Flow of sap between fruit and stalk is a part of the growth process. The amount of sap exuded varies with cultivar, maturity and location. As the fruit matures and ripens, sap flow decreases and eventually stops. Formation of an abscission layer blocks further sap flow and the fruit falls from the tree. If this sap comes in contact with the peel of the fruit, it causes a darkening or browning of the peel in the region of contact. This discolouration of the peel is referred to as sap-injury or sap burn-injury. Sap-injury in mango fruit is a major industry concern, as it results in poor fruit quality and lower prices, especially in the export markets.

Factors influencing sap-injury

Though sap-injury is a superficial damage, it reduces consumer acceptance and decreases shelf life of the fruit. Sap-injured skin could be invaded by *Aspergillus* spp., especially in hot conditions. Thus, it is a very important post-harvest handling and marketing problem of the mango industry, particularly in the case of fresh fruit. Sap obtained from the fruit harvested in the afternoon (2.00 pm) caused more sap-injury than the fruits harvested in the morning (7.00 am) and this was attributed to a dilution of injurious components of sap at higher fruit turgor. And also 'spurt sap' causes more injury than 'ooze sap'. Thus, the susceptibility of mango fruit to the injury as well as the severity of the injury seems to vary depending on the cultivar. Resorcinolic compound like 5-(12-heptadecenyl)- resorcinol, which is present in mango was also responsible for causing sap-injury.

Management

a) Physical methods

Different practices are followed in different parts of the world to control sap-injury in the harvested mangoes. In certain parts of India, the mangoes are de-sapped for 15-30 minutes on bamboo structures. In some parts of the world including India, one of the common practices is the inversion of fruits in soil for 15-30 min after harvest.

b) Chemical methods

In order to control sap-injury in mango fruits, several methods have been suggested.

- Plugging the stem-end of the fruit with chemicals after de-stalking, washing the fruit with 1% sodium bicarbonate or 1% aluminium potassium sulphate or applying surface coatings to the fruit prior to de-sapping are a few of the methods that are followed.
- Washing the sap-contaminated fruits with 1% sodium bicarbonate and 1% aluminium potassium sulphate to control sap-injury.
- Paraffinic oils to pre-coat the fruits that are harvested with stalk, before de-sapping.
- Subjecting the freshly stained mangoes to hot water treatment (51-55°C, 10 min) minimized the sap-injury; whereas subjecting the fruits to this treatment after the sap had dried.

i) Mango Decline

Mango decline is becoming a serious problem in Northern India. Old plants show the symptoms of declining trend invariably. Initially, the symptoms appear on older leaves with the collapsing of tissues. Besides, striking irregular brown necrotic area may also develop near the leaf tip and margin. Later, with increasing severity, almost the entire lamina gets affected. The younger leaves remain unaffected up to 2-3 months after their emergence. It gradually leads to the mortality of upper twigs progressing further to larger branches. Affected trees have thin canopies of lusterless green leaves and sickly look. The amount of chloride in scorched leaves was found to be more than that of the healthy leaves.

Management

- Adequate micronutrients application controls the disorder; mainly copper sulphate (1%) and boron (0.3%).
- Reduce the salt concentration to nontoxic level by heavy irrigation with fresh water.

j) Salt Injury

The dying back of the leaf tip and areas adjacent to the centre of the leaf exhibits an irregular area of dead brown and dry tissue which may sometimes involve maximum one-half of the area. Such an area is subsequently attacked by fungi. The cause of the tip burn is not known but salt sprays and salt in the water used for overhead irrigation causes one form of tip burn. Other factors viz., deficiency of moisture due to drought, high contents of salts in the soil and damage to root systems are the major contributing factors.

Management

- Growing of dhaincha as green manure in the inter space of the orchard.
- Application of gypsum @ 50 kg per tree of 50-80 years age around the basin of the mango is advocated.
- FYM and compost should be applied every year in the orchard.

k) Fruit Size Reduction

This abnormality was recorded in cv. Chausa at Saharanpur, Uttar Pradesh. In this abnormality, the fruit size and the shape is deformed as the lower half of the fruits is abruptly reduced in size. This abnormality may be due to the presence of wood rotting fungi on the twig portion.

Management

- Plant should be supplemented with proper micronutrients and fertilizers.
- Regular spraying of copper oxychloride (0.3%) on twigs and branches should be done twice in a year.

l) Fruit Tumour

Symptoms of this disorder are confined to fruits of polyembryonic. Symptoms on fruits include protuberate growth, starting near the tip and occasionally near the shoulder extending up downwards. It increases in size and becomes pale in colour with the advancement of disease. Surface of such affected area is scaberulous, which gives an ugly look. At acute stages, the protuberances (outgrowths) became necrotic, followed by deep cracks in the skin. Exudation of gum from the cracks is common phenomenon of the disease. Such necrotic skin turns brittle and dark brown and sloughs off easily in small chunks exposing the mesocarpic tissues. The flesh becomes musty, decayed and drops off prematurely.

Management

Use of micronutrients and recommended fertilizer is beneficial.

m) Chimeras

In this abnormality, portions of the fruits / leaf had distinct colour differences. Portion of the fruit becomes albino. Hence, distinct white and green portions are visible from the distance. Only few fruits in a plant show this type of abnormality, while other fruits remain apparently quite normal. This abnormality may be due to mutation. Such mutation occurs in part of a tree and produces fruits that are abnormally shaped, variegated and lacking in chlorophyll.

n) Fruit Atrophy and Deformity

Fruit deformation at an early age appears to become more severe. It has generally been noticed in Deshehari. The deshappened appearance of mango is the result of genetical defect influenced by environmental factors. Adjacent unaffected tissues continue to enlarge, imparting a clearcut deshappening, giving an uneven appearance to the fruit. Stones of affected fruits are deshappened according to the shape of the fruit. They are round, fork-like, pyriform, triangular, etc.

o) Sun Burn / Scald

Mango fruit may occasionally suffer major losses from sun burn. Damage generally occurs due to following factors:

- Very high temperatures that cause sun burn on exposed fruit on the tree.
- Fruit suddenly being exposed to sunlight when branches are broken.
- Holding harvested fruit in direct sunlight or left on the ground during summer.

Sun burn produces bleached or yellow patches on the surface of fruit. In severe cases, the skin is leathery, yellow brown to black, hard and slightly depressed.

Management

- Avoid any tree damage which may expose fruit to the sun.
- Do not leave fruit in direct sunlight or on the ground after harvest.
- Protect fruits from scorching sun during dry weather.

p) Jelly Seed

Jelly seed affects the interior of the mesocarp all around the endocarp and is characterized by softening of tissue around the stone. It is identified by a more intense yellow colour of the affected area compared with the rest of the mesocarp, which remains whitish or pale green in young fruit. This disorder shows no external symptoms except for a possible softening around the beak. As the symptoms develop, the yellow colour intensifies and reaches a larger portion of the mesocarp. The affected area eventually turns brown and softens to the point of having the consistency of jelly.

Management

Early harvesting of fruit is recommended to overcome the incidence of this disorder. Application of dolomite in acidic soils reduces the incidence of jelly seed.

q. Soft Nose

This physiological disorder is characterized by an internal breakdown of the mesocarp on the ventral side and towards the apex during the fruit development phase. Soft nose-affected fruits do not show external defects but yellowing of the green skin between the apex and stigma point or break with softening of the pulp occurs in some cases. The symptoms of soft nose are noticed only in fully mature fruit. The first sign of the disorder occurs next to the endocarp at the fruit tip, which spreads out into the rest of the apex and from there into the other parts of the mesocarp. The mesocarp undergoes a rapid localized over ripening at the fruit tip with the flesh eventually becoming watery and discoloured. The border between the disordered and healthy tissues appears clearly with the disordered tissue having a translucent appearance. While the pulp around the shoulder and the dorsal side remains unripe, internally affected flesh appears overripe but can be brown with a bitter taste. In severe cases, the mesocarp becomes a spongy and greyish-black mass.

Management

Maintaining low nitrogen during fruit growth and the sufficient level of calcium in the soil either by the application of calcium nitrate or by adding limestone or gypsum to soil. Early harvesting of fruit at 75% maturity is also recommended of mango fruit becomes prematurely pulpy and turns dark brown. The disorder is observed in fruits nearing maturity. The affected fruits do not ripen in storage, leading to an estimated loss of around 25%.

r) Girdle Necrosis

The first symptom is the disfiguring in the lower half of fruit with etiolated spots and appearance of brown dotted etiolated area. The brown dot-like portions coalesce and enlarge to form necrotic lesions. The lesions grow further to form necrotic girdles of tissues around the sinus region of the fruit, leaving the green apex healthy in the initial stages. In the later stages of the disorder, the tip turns necrotic, closely resembling the symptoms of black tip. The seed protrudes from the pulp with layers of decaying mesocarp tissue attached to it and in rare cases, a whitish brown gummy substance exudes from the necrotic tissue before the epicarp shows discolouration. The mesocarp tissue disintegrates forming a cavity underneath the necrotic spots that enlarges and becomes continuous below the epicarp in sinus region of the fruit. The seed is covered by a thin layer of necrotic mesocarp tissue and leading to girdle necrosis.

Management

- The best method of controlling the disorder is to locate the orchards at least 1.5 km away from brick kilns. In those cases, where a brick kiln exists close to the orchards, adequate measures must be taken to prevent the fumes from reaching the orchards during the entire season or at least during the fruiting season. Practices that are possible include avoiding kiln operation between flowering and fruit maturation, raising the height of chimneys to 15m or more so as to reduce the effective concentration of harmful gases in the vicinity of tree canopy and using smoke filters on chimneys.
- Chemical methods of control could also be employed where necessary by preharvest spray application of 0.6% borax, 0.5%

washing soda, 0.8% caustic soda, or a mixture of 0.6% borax and 2% sodium carbonate. The first spray of borax is given at pea stage followed by two more sprays at 15 days intervals.

s) Chilling Injury

Chilling injury in mango is among the most damaging disorders as the fruit is known to be susceptible to chilling injury at storage temperatures less than 12.5°C. It is an economically important post-harvest problem that reduces the overall quality and marketability of mango and occurs in transit or distribution, in retail or storage. Chilling injury is defined as the damage resulting due to exposure of fruit to low temperatures above the freezing temperatures of mango (-14°C). It is the manifestation of several physiological and biochemical changes occurring during storage at low temperatures and expressed after transferring to non-chilling temperatures. It is distinct from freezing injury caused due to cellular dehydration following extracellular freezing of water and damage from the ice crystals that develop subsequently. The most common visual symptoms of chilling injury in mango are dark skin, scald-like discolouration and pitting or sunken lesions on peel beginning around lenticels and spreading outward to produce a more or less circular lesion, development of off-flavour, discolouration of the pulp and overall poor fruit quality. Other symptoms include uneven ripening, poor colour and flavour, increased susceptibility to decay and in severe cases flesh browning.

Management

- Reduction in chilling injury may be obtained by pretreatment of fruis with 2,4-D, polyamines, ethylene and methyl jasmonate.
- Chilling injury in ripe mangoes can be alleviated by holding preclimacteric fruits at 10°C for a minimum period of 30 days followed by ripening at 27-34°C.
- Chilling injury can be avoided by storing the mango fruit above its critical temperature, which varies widely depending upon the cultivar.

t) CO_2 Injury

Exposure of mature green mango to oxygen levels below 2% and/or carbon dioxide levels above 10% for longer than a few days may induce skin discoloration, greyish or pale flesh colour, uneven ripening and off-flavour development due to fermentative metabolism with the accumulation of acetaldehyde and ethanol. Ripening and senescence rates in mango are affected by the level of O_2 and CO_2 in the storage atmosphere during respiration and show a significant inhibitory effect on ability of ethylene to initiate ripening. Carbon dioxide injury is also associated with the development of internal pitting and browning with associated exudates and inhibition of ripening.

Management

CO_2 injury is best avoided by maintaining O_2 and CO_2 levels within safe limits.

u) Lenticels Discolouration

Lenticels discolouration is a superficial blemish that affect certain varieties of mango by exhibiting a speckled appearance making it less desirable. Blemish development is only superficial and does not penetrated in to the layer of the fruit. The development of a red halo is the first indication of the lenticel discolouration. In some cases, the redness intensify and ultimately results in a localized dark end appearance.

Management

Fruit bagging during fruit growth and maturation reduces lenticel discolouration.

❑❑❑

2

Citrus

a) Granulation

Granulation is a physiological disorder of juice sacs of citrus fruits, wherein, they become comparatively hard, dry and assume a greyish colour, become somewhat enlarge and have insipid taste. Such fruits fetch poor price in the market, as a result of which growers suffer heavy losses. It is more serious in northern parts of India than in southern parts. Citrus areas having a humid climate are reported to have higher incidence of granulation than dry areas. Sweet orange and mandarin suffer seriously from granulation, whereas it is less serious in other citrus fruits. It is interesting to note that granulation is generally not evident until the fruit is nearly or fully mature. Moreover, it is not possible to differentiate between the granulated and non-granulated fruits on the basis of their external appearance. On cutting only, one can differentiate between granulated and non-granulated fruits. In granulated fruits, the juice sacs are somewhat enlarged, hard, contain less orange colour, are lignified and have little running juice.

Rind colour of fruit in sweet orange cultivars change from orange to yellow, more prominently at the stem-end, where the intensity of granulation is maximum. However, granulation can now be detected without cutting the fruit by the laser doppler detection technique developed in Japan. During granulation, the vesicles develop gas bubbles, which results in increase in volume and a decrease in fruit weight, thereby, decreasing the specific gravity of granulated fruits. The specific gravity decreases from 0.93 of the normal sweet orange fruits to 0.64 in severely affected fruits with granulation.

Changes occurring during granulation

Normal juice vesicles in the first stage become slightly hard and enlarged due to increase in cell size near the stalk-end and tapering towards the free-end and contain viscous jelly like substances instead of running juice, high percentage of moisture, inorganic matter, polysaccharides and low amounts of sugars and acidity. The alcohol insoluble solids like cellulose, hemicelluloses, lignin, starch and pectin substances increases during the enlargement and hardening of the granulated vesicles. The cell wall of these hard juice sacs is thickened due to gelatinization and lignifications.

In the advanced stage, the granulated juice vesicles become dull yellow and later colourless due to oxidative degradation of carotenoids. The juice vesicles contain a cavity in the centre, which contains large amount of carbon dioxide, volatile substances and pectins. In the third stage, the central cavity is enlarged and subsequently, the vesicles collapse, give off large amounts of CO_2, alcohol and volatile substances presumably due to onset of the fermentation that disturbs the normal metabolic functions and accelerate the senescence and disintegration of the tissues. The total soluble solids, protein, neutral sugars (rhamnose, arabinose, mannose and galactose), soluble sugars (glucose, fructose and sucrose) organic acids and ribonucleic acid (RNA) decrease dramatically with increase in the intensity of granulation. The hardened cells undergo softening with progressive disintegration of the inner cells and their appearance is granular. Several factors have been found to be associated with it. They are as follows:

Temperature

Prevalence of high temperature has been reported to favour granulation.

Light

The role of light in granulation differs with the age of the tree and with the species as well. Fruits receiving full light have less incidence of granulation as compared to those grown under shade. Kinnow mandarin is almost free from granulation and one of the reasons attributed to it may be that it bears fruit inside the canopy.

Humidity

Citrus areas having humid climate are reported to have higher incidence of granulation than the dry regions and thus, the incidence of granulation is higher in coastal regions.

Growing season

The incidence of granulation is more during major fruiting season and less in winter. Similarly, its incidence is higher in "off" year than in "on" year.

Tree age and health

The degree of granulation does not differ much between young and old trees but the incidence of granulation is considerably higher in younger trees than in older ones. In case of old trees, tree health (healthy or declining) has no bearing on granulation. However, the extent of granulation is higher in the declining trees as compared to healthy ones.

Tree vigour

Conditions favouring luxuriant growth are associated with granulation. It's incidence is high in the fruits produced on old trees, which are heavily pruned in the previous seasons and produce luxuriant growth in the following season. Similarly, cultural operations like frequent irrigation and application of nitrogen favour granulation as these induce more vigorous growth.

Tree aspect

Northern side and interior of the tree have usually higher incidence of granulation than on southern side. The fruits located on the northern side of the tree have shorter day length than those on southern side.

Tree variation

Some trees within an orchard have been reported to show greater tendency to granulate, while the others produce healthy fruits. Further, the incidence of granulation in fruits of a tree seems to be persistent in the sense that trees producing granulated fruits would always produce granulated ones, while plants producing non-granulated fruits would always produce non-granulated fruits.

Fruit size

Larger fruits are more affected by granulation than the smaller fruits.

Harvesting time

Generally, the fruits harvested late in the season tend to show more granulation. The incidence of granulation usually appears during the later period of fruit.

Species and cultivars

Different citrus species and cultivars within the species vary widely with regard to the incidence and extent of granulation. Generally, sweet oranges are more prone to granulation than mandarins, tangelos and grapefruits. However, Kinnow mandarin, Minneola and Orlando tangelos and Duncan grapefruit are almost free from granulation.

Rootstocks

Rootstock is a vital component of citriculture. Rootstocks influence the incidence and extent of granulation, though their influence is not much consistent and also depends on climatic factors.

Nutrition

Malnutrition of the citrus plants may favour granulation in the fruits. In general, zinc deficiency is associated with granulation. The trees having the highest foliar nitrogen content produce the maximum granulated fruits. Low nitrogen and phosphorus, high potassium and boron and an excess amount of lime, which contains calcium also favour this disorder.

Management

Citrus granulation is a complex problem and several factors are associated with it. There is no conclusive information for its control. However, some remedial measures can reduce it's incidence to some extent. They are as follows:

Use of rootstock

As the vigorous rootstocks like Rough lemon and Sour orange favour granulation, less vigorous rootstocks may be tried.

Irrigation frequency

The amount and frequency of irrigation may be reduced. In general, irrigation at 14 days interval is the best to reduce the incidence of granulation.

Use of lime

Calcium hydroxide (two per cent) applied three to four times at monthly intervals starting from August reduces its incidence considerably.

Use of potassium nitrate

Four sprays of potassium nitrate (one per cent) starting from August onwards at monthly intervals reduce the incidence of granulation.

Spray micronutrients

Zinc and copper are the important micronutrients for the overall health and productivity for a plant. Two sprays of zinc sulphate (0.5 per cent) during September and October are useful for reducing the incidence of granulation to a greater extent. Copper sulphate is also very effective.

Similarly, four sprays of 25 ppm boric acid starting from August at monthly intervals has been found useful in reducing the incidence of granulation.

Use of plant growth regulators

Several growth regulators like Gibberellic acid (GA_3), Naphthalene acetic acid (NAA), 2, 4-dichlorophenoxy acetic acid (2, 4-D), Ethrel, etc. have been used very widely in different concentrations at different stages of fruit development. Further 2, 4-D (15 to 20 ppm) is highly effective for reducing the granulation. Similarly, three sprays of Planofix (NAA) at 300 ppm during August, September and October decreases the incidence and degree of granulation in many citrus types. Application of GA_3 between August and November at 10 days interval reduces granulation considerably. Ethrel (ethephon) at 500 ppm not only reduces the citrus granulation but also improves fruit quality.

Harvest at right time

Delayed harvesting usually accelerates granulation in sweet oranges and mandarins in India. The incidence as well as intensity of granulation increased if fruits are left on plant for longer period. Therefore, picking of fruits at the right time is equally helpful in reducing this incidence.

Precautionary measures

It is always useful to adopt some precautions to overcome the problem of granulation in citrus. These are as follows:

- Select material for planting from plants, which have low granulation history.
- Avoid heavy doses of nitrogenous fertilizers.
- Avoid heavy pruning. If necessary, follow light pruning.
- Ignore every cultural practice which favour luxuriant growth.
- Keep rigid control on time and frequency of irrigation.

b) Fruit Splitting in Citrus

Splitting or cracking is a common feature in acid lime, lemons and sweet oranges. It is a physiological disorder associated with sudden changes in weather conditions. This problem can be found in citrus orchards during the summer seasons. Cracking begins in the peel at the stylar-end of the fruit, although it may originate midway between the ends. Splitting may be of longitudinal or transverse. But longitudinal splitting is commonly found under normal conditions and in extreme cases, there is a complete splitting of the fruits. Splitting or cracking occurs after rains following a drought. High temperature during the drought period makes the rind of the fruit hard and less elastic. During the rains, the growth processes are accelerated, the tissue inside the fruit expand. The peel seemingly fails to enlarge as fast as the pulp and therefore, cracks. Sometimes hot winds and defective irrigation practices lead to splitting of fruits.

Management

- Application of recommended level of potassium fertilizers.
- Frequent and light irrigations during the hot and dry period with an interval of 6 to 8 days.
- Timely picking of fruits so as to reduce losses due to fruit cracking.

c) Citrus Decline

Citrus decline, often referred to as citrus "die-back" is not a specific disease but is a symptomic expression of many disorders in the plant. The malady is also named as 'frenching'. 'chlorosis', etc. Terms like die-back or decline are now used for certain specific disorders in other countries. The term "die-back" is usually used as a synonym for "exanthema" of copper deficiency. The term actually signifies the continuous dying of twigs.

Factors responsible for citrus decline

Soil Factor

The presence of any impregnated layers of calcium carbonate, or clay pan may affect the permeability and aeration of soils as well as the root growth and drainage. The depth of kankar or clay pan in the soil determines the depth of active root system and also the age at which the decline starts. Decline is delayed by the increasing depth of hard pan. Bad drainage causes poor aeration. The problem of poor drainage exists in many parts of the country. Under the conditions of poor aeration, the root development is arrested, leading to less absorption of nutrients and ultimately to decline of the top.

Citrus trees are quite sensitive to excess of salts and are injured, if the total salt concentration in the soil exceeds 1000 ppm. The presence of excessive salts in any of the horizons up to a depth of 180 cm adversely affects the health of citrus plants. High $CaCO_3$ content in soils is an important factor for the decreased availability of iron and zinc in soils. Deficiencies of zinc, iron and manganese are quite common on calcareous soils. A high pH not only affects the availability of micronutrients, but a high sodium content of such soils may have direct toxic effects in the case of sensitive crops like citrus. A low pH allows the leaching of nutrients. Excessive irrigation is injurious for the health of citrus plants.

Nutrition

Decline of citrus trees in South India is due to lack of adequate plant nutrition. Sweet oranges in South India suffer from deficiencies, the most common on the plants being of zinc in some cases, lack of copper may cause trouble and in acid sandy soils other elements may also be lacking. The excess of iron and salinity and high uptake of phosphorous and manganese are the causes of die-back have been pointed out.

Rootstock

Rough lemon as rootstock has shown incompatibility with Mosambi (sweet orange) in Maharashtra with Sathgudi in Andhra Pradesh, Blood Red and Mosambi in Punjab. In many instances, failure of citrus trees may be traced to incompatibility between stock and scion. In Maharashtra, Rangapur lime as the rootstock for Nagpur oranges was found to be better than Jamberi. Susceptibility of rootstocks to root rots, salinity, waterlogging, cold, viruses, etc. also leads to citrus decline.

Orchard management

In many cases, decline of citrus trees can be traced to mismanagement and negligence. Excessive intercropping, faulty, excess or no manuring, poor quality of irrigation water, are some of the common causes leading to citrus decline.

Insect-Pests

A large number of insects attack citrus trees. Insects cause damage to citrus in two ways viz. (i) by direct feeding and (ii) by acting as vectors for the transmission of viruses. These are the principal contributing factors of citrus decline. Incidence of diseases in citrus is high. Several diseases contribute towards citrus decline.

Nematodes

The incidence of nematodes has been recognized as one of the contributing factors of citrus decline. They cause damage to citrus trees by feeding on roots. Nematodes were found to associate with declining trees.

Fungal diseases

Incidence of diseases in citrus is high. Several diseases of citrus contribute towards citrus decline. Survey of the citrus orchards in different parts of the country revealed that *Colletotrichum gloeosporioides* is predominantly associated with the die-back.

Viruses

Viruses are the most potent factors leading to citrus decline. The presence of tristeza reported from every state in India. Other viruses contributing to citrus decline are psorosis, xyloporosis, stubborn, exocortis, etc. Virus infected trees do not put on normal healthy growth, become stunted, show dieback of twigs and branches. The productivity is substantially reduced. Another important contributing factor is greening mycoplasma. It is reported from all over the country.

Symptoms of citrus decline

- Symptoms vary with the cause of the malady.
- All the citrus species and varieties are susceptible, but the sweet and mandarin oranges are worst affected.
- Budded, as well as, seedling trees are highly susceptible.
- The trees do well in the first few years of their orchard life, bearing very little crop and ultimately become uneconomical.
- They show symptoms of ill health and decline.
- The affected trees do not always die completely, but remain in a state for decade and unproductive for a number of years.
- Sometimes, they may suddenly wilt and die in a day or two.
- In early stages, symptoms are restricted to a few limbs, but eventually the whole tree is affected.
- Trees show sparse mottling leaves, stunted growth, sickly appearance. midrib and lateral veins of old, mature leaves turn yellow with interveinal areas along the veins showing diffuse yellowing.
- Leaves may turn yellow and are shed with the onset of summer or autumn and the die-back of twigs starts.
- Dead shoots stand out prominently and may be found dead right down to the main trunk.
- The entire tree bears short twigs carrying narrow small leaves on their lower portion. Subsequent secondary growth consists of short, upright small, weak shoots showing a variety of discolouration of leaves.
- Often these leaves have green veins of green blotches. Occasionally, small, circular, green spots appear on yellow tissue on leaves. The die-back of weak shoots continues.

- There is excessive flowering, but the fruits are not carried to maturity. The fruits show distinct sun-blotching.
- The feeder root system becomes depleted, roots turn black and sometimes are covered with rotting bark.
- Either only a few trees or entire orchard may be affected.

Management

Good cultural practices, improvement in soil fertility and drainage, control of insect pests, nematodes, etc. may be useful to minimize the incidence of decline. Use of resistant rootstocks and certified bud wood for propagation is also useful. Spray the tree canopy with a cocktail of micronutrients during April - May every year.

Zinc sulphate	100 g
Copper sulphate	60 g
Magnesium sulphate	40 g
Ferrous sulphate	40 g
Manganese sulphate	40 g
Lime	180 g
Water	20 litres

In acidic soil with pH 4.5-5.0, apply 100 g of Dolomite lime/plant once in 3 years

d) Irregular Flowering

The soil properties, such as high clay content, mainly concentrated in sub-surface, low free lime content, coarse fragments and sand content were found conducive for irregular flowering behaviour of Nagpur mandarin. These properties account for lesser moisture depletion during water stress period, with the result, the required quantum of stress is seldom obtained and plants flower erratically.

Management

The only remedy possible in irregular flowering in orchards is to stop the vegetative growth with the help of growth retardants which otherwise continue due to constant moisture supply from sub-surface layers. Application of Paclobutrazol @ 18g/plant at the time of imposition of water stress is found very effective in inducing good flowering.

e) Frost Damage

Leaves, twigs look water soaked, then wither and turn darken. Leaves may drop quickly or persist on tree. Flesh of fruit dries out and brownish pits (ice marks) may form on the rind. Branches die back and bark splits in severe cases.

Management

Allow the tree to recover before removing frost-killed wood. After new growth appears in early spring, wait for any die back, then cutback to live wood (identified by a green layer just under the bark). Pruning cuts will heal naturally, so there's no need to paint them.

f) Over Watering

Leaves yellow and drop. Abnormal number of blossoms.

Management

Decrease irrigation frequency. Avoid planting ferns, annual flowers, or plants that need lots of water near citrus trees.

g) Fruit Drop in Citrus

Yield of citrus fruits is erratic and is not consistent during all the years of their bearing life. Heavy fruit load followed by low or negligible and medium crops is a common phenomenon in citrus, though the trees are apparently healthy and supplied with judicious manuring. Enormous shedding of fruits occur abnormally right from fruit set until maturity and reduce the potential crops resulting in an uneconomical cultivation particularly during off years. The fruit drop in citrus is serious as with apples. Heavy as well as poor set coupled with high rate of fruit drop distresses the fruit growers. The retention of the fruit depends upon many factors both internal and external.

Fruit drop is high due to high temperature and dry atmosphere. When there is stress of water in soil or plant, fruit drop is also more. High wind velocity, hail storms cause considerable fruit drop. Insect pests viz., mites, bugs, etc. also cause heavy drop of fruits in citrus. Similarly, diseases cause fruit drop. Besides, physiological factors which help in formation of abscission layer at the stem-end point cause fruit drop. Imbalance of certain hormones, such as auxins, cytokinins, gibberellins, etc. or deficiency or excess of essential nutrients may cause fruit drop. Genetic make up of the species and varieties also affects incidence and intensity of fruit drop. As a natural rule, all the blossoms do not set fruit even though conditions are ideal for their maturity. Immediately after initial fruit set, enormous drop of young fruits occur. In spite of the natural fruit drop, sometimes a reasonable number of marketable sized fruits is obtained at harvest. This is known as normal fruit set.

Periodicity of fruit drop

All the fruits that fail to reach maturity do not drop at one time and continuous dropping from flowering stage upto maturity is not common. Instead, there are more or less definite periods/phases/ waves of accelerated or excessive fruit drop when extensive dropping occurs. These are designated as waves, stages or spells of fruit drop. The loss comes in a series of waves varying with the different fruits in number and length of time between them. There appears to be certain critical periods through which each fruit must proceed to reach full maturity. When one of these critical periods are safely passed, there is comparatively little danger of fruit falling before the next critical period arrives. Following are the waves of drops with special reference to mandarin:

(i) Post setting or immature fruit drop

This is a drop of young button fruits which take place immediately few days after a petal fall. These dropped fruits have calyx attached. This wave of fruit drop occurs due to auxin deficiency caused by rapid cell division and generally takes place after fertilization till endosperm attain its growth.

(ii) Summer drop

This is also known as the third drop. During this period, fruits of ber size drop down. Dropped fruits appear externally normal but are without persistent calyx, softened and lack in turgidity. The cause of this fruit drop has been reported due to competition for food material among the large number of fruits held on the tree and also is favoured by high atmospheric temperature, accompanied by low humidity coupled with summer months at the time.

(iii) Premature fruit drop

This kind of drop occur prior of fruit maturity and dropped fruits are of 1/2 to 3/4 growth in size and do not show any symptoms of internal or external injury but comparatively smaller in size having yellowish colour near the stalk end as against the fruits retained on the tree, that are completely green and relatively of bigger size. The causes of this drop are excess soil moisture during rainy season and embryo abortion, which leads to auxin deficiency and formation of abscission layer.

(iv) Preharvest drop

In citrus, during maturation process, rate on fruit drop increases rapidly till harvest. The fruits abscissed during this preharvest period have yellow to orange colour. Declining rate of auxin content due to seed maturity and embryo abortion is the cause of abscission.

Management

- Apply recommended doses of manures and fertilizers.
- Irrigate citrus crop lightly at frequent intervals.
- Avoid citrus planting in the areas, where water table is high.
- Spray 2, 4-D @ 5-10 ppm twice at marble stage of fruit development and repeat the same at pre-mature stage.
- Apply micronutrients and take appropriate plant protection measures carefully.

h) Frost Injury

Frost injury usually occurs after nights with clear skies and little wind causing temperatures below minus 2.2°C for more than 4 hours. Damaged fruit may show drier juice sacs and separated segments without injury to the rind.

Management

- Selection of warmer planting sites (elevated ground where cold air can drain away).
- Selection of more cold tolerant scion/rootstock combination.
- Wrapping the trunks of young tree.
- Supplying sufficient water to the roots and turning on overhead reticulation during periods of frost.

i) Oleocellosis

Oleocellosis is a skin injury caused by oil released from the oil glands in the skin after they have ruptured. The damage can be caused by physical means (poor handling or abrasion) or climate. The symptoms are greenish or brownish, firm, irregular patches or spots on the skin. Damaged areas typically darken and become sunken with time. The disorder is sometimes called green spot because the affected areas do not colour at ripening.

Management

- Oleocellosis can be avoided by careful handling and avoiding harvesting fruit early in the morning or after rain.

j) Wind Injury

In areas with persistent winds, fruit on the outside canopy can be scarred where twigs or thorns rub against the rind. Fruit can also drop from the tree. Strong, persistent wind causes water stress by dehydrating the leaves, causing leaf necrosis or twig dieback or stunting of the growth of young trees.

Management

Appropriate irrigation and windbreaks can help reduce wind damage. Individual shelters of wood frames may be suitable for young trees.

k) Sunburn

Heat produced during direct exposure to intense sunlight can blemish fruit, cause chlorotic or necrotic leaf blotches and produce bark cankers. On sun burned fruit, the side exposed to the sun develops brown, leathery areas. Some fruit become slightly lopsided.

l) Rind Disorder

A rind disorder that develops on mandarins sometimes results in severe crop loss. Brown, water-soaked blotches or dark, sunken areas develop on exposed surfaces of fruit in the outer canopy. Fungi develop in the discoloured areas and the flesh underneath softens and rots. Mandarin rind disorder occurs during high rainfall season. Once fruit begin to develop mature colour, rind injury becomes apparent within days after rain. The disorder is a physiological problem that develops during adverse environmental conditions, when the rind becomes water soaked and the oil glands rupture. The injured rind is then colonized by various secondary decay fungi.

Management

Treatments include one or two foliar applications of oil or antitranspirants that repel water, applied before forecasted rain or around fruit colour break.

m) Puff and Crease

The innermost spongy white layer of the rind sometimes separates from the outer surface of the fruit segments. This internal damage is visible externally as an uneven appearance of the rind surface. Some portions of the rind surface appear inflated (puffy) and other areas are indented (creased). Hence, the name "puff and crease." Navel and Valencia oranges are susceptible to this tissue separation.

Symptoms become increasingly prevalent as fruits mature. Puff is sometimes considered a separate malady from crease, but their overall damage appearance is similar. The cause of rind separation from flesh and how to prevent it are not well understood.

Management

The application of plant growth regulators and various cultural practices may reduce this problem.

❑❑❑

3

Guava

a) Die-back

This disorder affects young as well as the old bearing trees. Some trees are more prone to attack. The typical symptoms include withering of lowermost branch of the tree from top downwards. The growing tip turns dark brown and necrotic areas extend backwards. A typical lesion develops at the junction of the diseased and healthy areas which advances down the healthy areas. The infected branches defoliate giving them a barren appearance.

Management

Application of 1.8 kg lime or gypsum will reduce the mortality of trees by maintaining soil pH.

b) Fruit Cracking

Cracked fruits deteriorate rapidly and often invite secondary infection by disease causing microorganisms and insect-pests. Consequently, fruits become unmarketable and ultimately cause great economic loss to the growers. In guava, generally radial cracking has been noticed. In this type of cracking, surface of the mostly full ripe fruits crack radially from the distal to proximal end. The mechanism behind cracking is that the strengthened tissue that was developed in xylem and phloem loses their ability to divide and enlarge. Another reason is, increased water supply after dry spell. Due to these reasons, meristematic tissues quickly resume growth but not the strengthened tissues. Owing to differential growth rates, hardened tissues rupture and some incipient cracks originate at hypertrophied lenticels. It has been promoted by greatly retarded transpiration, accompanied by plentiful water supply to the regions of hypertrophy.

Fruit cracking in guava may be associated with various factors:

- Environment - Hot winds, sharp fluctuation in day-night temperature coupled with heavy irrigation after dry spell and high relative humidity leads to fruit cracking.
- Hormone - Low auxin content in the seed is directly responsible to fruit cracking.
- Nutrition - Nutrients like potassium, zinc, boron, manganese are involved in physiological processes during fruit growth period and thus, their deficiency may result in fruit cracking.
- Soil moisture-Increased water supply, when water loss through leaves gets reduced due to high relative humidity, causes fruit splitting. Defective irrigation practices also may result in fruit cracking in guava. Heavy irrigation at fruit ripening stage after the long period of dryness leads to fruit cracking.
- Diseases, pests, mechanical injury rupture of skin at the weakest point because of physiological disorder or damage by disease or insect or mechanical injury.

The ruptured tissues when exposed to atmosphere, deteriorate rapidly. Thereafter, it suffers secondary infestation by invasion of disease causing organism, such as *Fusarium, aspergillus* and insect-pest leading to partial rotting and early premature dropping of fruits.

Management

- Pre-harvest spray of plant growth regulators, such as NAA, GA and ethrel for reducing cracking. These chemicals suppress the ageing process and also increasing the plasticity and elasticity of fruit skin.
- Grow tolerant type of cultivars.
- Foliar spray of potassium, boron and copper increases the fruit growth rate by stimulating enzymatic action in the peripheral tissues.
- Follow basin system of irrigation and ensure regular supply of water to each tree in the orchard.

c) Bronzing

This is the most important problem affecting guava production and quality and life of the trees. In Southern India, the old leaves turn purplish, the fruits become small and hard. On moderate and severe cases, the twigs die back from the tip, the entire leaves become bronze coloured, the bark of trunk near ground splits and the tree dies within 2-3 years.

Management

Since this malady is brought about by a deficiency of a combination of nutrients, the following micro nutrients has to be applied through foliar spray. Boric acid, zinc sulphate and copper sulphate, each of 300 g along with 100 g of urea and 100 ml of soap solution dissolved in 200 litres of water will be sufficient for an area of one acre. Basal soil application of 30 g each of borax, zinc sulphate and copper sulphate has to be done around every tree.

❑❑❑

4

Banana

a) Seediness in Banana

Poovan (AAB) is by for the leading commercial variety of banana occupying more than 70% of the area under banana cultivation in India and manifest a peculiar developmental disorder - a predisposition to seediness, locally known as "**Kottaivazhai**". This malady is characterized by the presence of distinctly conical and ill-filled fruits with a prominent central core having many under-developed seedy structures rendering the fruits inedible. It has a potential to reduce the production by 10-25%. Because of the deceptive nature, this disorder poses a serious threat to engulf the entire area under poovan imperceptibly and the banana growers are getting bewildered. This malady could be identified at the later stages of development from the appearance of streaks, striations and blotches on the pseudostems and on the leaves.

This is also genetically inherited, as there is a presence of heavy and intense blotches and striations even on the second crops of "Kottaivazhai". The bunches are held at an angle above the horizontal position. The peduncle showed irregular curvature and exhibited a shape. Pollen of kottaivazhai is infertile, shriveled, shrunken and broken. The seeds are soft, flat, ill-filled and regularly angled. There is not much difference in the basic structure of the fruits of normal poovan and kottaivazhai (Table 1). The area of the pericarp in kottaivazhai is much smaller than that of normal poovan. The locular cavity of kottaivazhai is bigger than that of normal poovan fruits. Seed formation occurs during the fourth week after formation.

Table 1. Comparison of developmental stages of kottaivazhai

S.No.	Weeks after flowering	Normal Fruit	Kottaivazhai
1.	4th	Ovules are disorganized, fruit increased length and diameter considerably. Outer epidermis of the pericarp keeps pace by the enlargement of its cells in the tangential direction. Ovarian cavity disintegrates.	Seed formation occurs. After this, ovarian cavity does not disintegrate. Ovules still grow and the funicle becomes prominent. Slight increase in fruit volume due to increase in size of seeds.
2.	8-12th	Floral axis and septa greatly expands. Ovule is organized.	Not much difference in growth, but the seeds enlarge in size.
3.	17th	Entire ovarian cavity is completely filled with starchy parenchymatous tissues; ovules are disintegrated.	Little growth of pulp; but it could not completely cover the locular cavity; seeds size attain the maximumat this size; However, in the bunches treated with PGRs, the ovarian cavity is completely filled up at this stage without the formation of seeds.

Management

Spray of both Gibberellic Acid (50 ppm) or 2,4-D (20 ppm) immediately after the last hand is opened, the seediness in kottaivazhai is completely overcome and the treated bunches are found better than normal poovan in respect of the bunch and fruit characters and quality.

b) Abnormalities in Banana

An abnormal case of a banana plant producing two flower buds from single stalk. A plant with five bunches has been observed. In a perennial plantation at Aduthurai, Tamil Nadu, a plant of the var. poovan with two bunches emerging from the crown was observed.

The pseudostem of the plant was robust with a height of 230 cm and girth of 58 cm. The plant characters recorded were identical with that of the normal poovan variety. The forking of the inflorescence axis into two distinct peduncle was noticed just at the point of emergence below the flag leaf. When the pseudostem was split open to reveal the entire length of the inflorescence axis, the time of differentiation of the two peduncles was visible just from the scar of the leaf sheath from which the flat leaf has emerged. The two bunches were very nearly similar in appearance and were less than normal size. A common sterile bract was noticed with distinct foliage character, emerging from well inside the crown of the pseudostem, before the forking of the inflorescence axis.

c) Goose Flesh

Goose flesh is a physiological disorder, which affects the banana peel. The symptoms of this disorder are observed on ripe fruits and the degree of damage varies from slight to severe. Affected fruit has a wilted and shriveled appearance and the peel usually turns brown, when the fruit is handled. Limited handling possibilities as well as unattractive appearance render "Goose Flesh" fruits unpopular, fetching low market value. The Goose flesh peel disorder occurs more frequently during the dry winter, when the atmospheric humidity fells below a relatively low level.

d) Yellow Pulp

Abnormal growing conditions, nutritional imbalance, stress and defoliation can result in yellowing or a honey-coloured tint to the normally dull white pulp at harvest. This is often accompanied by a softening of the pulp in the centre of the finger. Premature ripening, usually in transit often follows yellow pulp symptoms. Factors that delay fruit maturation, such as severe leaf spot, excessive shade and drought can cause yellow pulp. Magnesium deficiency can also result in delayed maturation and yellow pulp.

Management

Magnesium and sulphur application reduces the incidence.

e) Comb Effect

Occurrence of this order is due to the magnesium deficiency in the plantation. In this disorder, chlorosis is marginal at first, sometimes with a tiny green edge remaining on the leaf margin. It spreads towards the midrib but more quickly to the main veins than to the interveinal areas, which remain greener from a long time giving a Comb Effect. These symptoms affect the entire plant, it's foliage turning a dirty yellowish green.

Management

Foliar spray of 1% magnesium sulphate reduces the disorder.

f) Degrain of Bunch in Banana

This disorder is characterized by the dropping of ripe banana fruits from a bunch due to rotting of the pedicels. Long stalks are mechanically weaker than the shorter ones and they also differ anatomically. High N/P ratio in the manure tends to thicken the stalks and flatten the central cylinder and markedly increases the susceptibility to tannin oxidation.

Management

Enhanced application of phosphorus (preferably DAP application) would alleviate the disorder.

g) Peel Splitting

Longitudinal split of the peel, usually beginning from the proximal end near the pedicel. The split usually divides the peel into unequal halves and ultimately expose the pulp as the split widens.

h) Chilling Injury

Chilling injury is the term used to describe physiological damage to fruit tissues resulting from the exposure of chilling-sensitive fruits to temperatures below a critical threshold. In banana, Chilling injury can occur in unripe or ripe fruit and may reduce fruit quality and market value. Chilling injury avoidance represents a key economic step in the banana food chain. Indeed, this physiological problem is only evidenced during fruit maturation and each storage place particularly after ripening induction (e.g. at the market place) must clearly avoid a storage under this critical temperature (i.e. 13°C for Cavendish). This highly sensitive fruit should be definitely be stored separately to the other non-sensitive fruits stored at very low temperature. Surface lesions, such as pitting, large sunken areas and discolouration of the surface, dark water soaked areas of the peel, internal discolouration (browning) of pulp, breakdown of tissues, failure of fruits to ripen normally are the normal symptoms.. When chilling is severe, fruits develop extensive sub-epidermal browning and eventually turn black. There is a loss of the development of the characteristic flavour, aroma and taste and often the development of off-flavour.

Management

- Chilling injury can easily be avoided in banana cultivars by limiting storage or handling to temperature above the threshold.
- Avoid refrigeration at below 13°C (55°F).

i) Choke Throat

Choke throat occurs, when the bunch is about to emerge from the top of the pseudostem but becomes trapped at various stages of emergence. In its most severe form, the bunch fails to emerge from the top of the pseudo stem and instead bursts through the side of the pseudostem. Such bunches are worthless. In severe cases, the top one or two hands become trapped in the throat of the plant leading to badly misshapen fruit. Bunches which experience difficulty in emergence usually have several leaves closely 'bunched' together at the top of the pseudo stem. Choke throat is seasonal in nature. It is usually in the winter and early spring, following cold weather. It can also occur in waterlogging or severe water stress conditions and following wind storms.

Management

- Select taller varieties, which are less susceptible to choke throat.
- Choose a warm environment, one which is well protected from frosts and strong winds. Slopes facing the North and North West are usually warmer.

- Control time of bunching to avoid cold weather prior to bunching. Plants bunching in the late spring to mid autumn are less affected.
- Regular irrigation to avoid water stress particularly during hot-dry weather.

j) Hard Lump

Rasthali is a very delicious and dessert variety. The disorder of hard mass or lump commonly occurs in this variety. Lump are pinkish brown in colour, less luscious and firmer than the usually soft pulp and tastes like immature or unripe fruits, resulting in impaired eating quality of the fruit. Lump development is related to lower day temperature during fruit development period.

Management

- This disorder can be overcome by adjusting planting time so as to harvest bunches during January-June.
- By treating cut end of the harvested bunch or hand with 2,4-D at 500 to 1000 ppm.

k) Neervazhai

This disorder has been noticed in the 'Nendran' banana grown in Tiruchirapalli district of Tamil Nadu. Very low percentage of plants is affected and suckers from affected clumps carry the disorder. It is characterized by fruitless bunches and cannot be detected until bunch emergence i.e. no other symptoms appear. The bunches have 4-5 hands and have only neutral flowers, the remaining nodal clusters consist of only persistent male flowers. This malady may be due to imbalance of hormone or hormones influencing female flower development.

l) Abrasion Injury

Abrasions result from skin scuffing against other fruits or surfaces of handling equipment or shipping boxes. When exposed to low (<90%) relative humidity conditions, water loss from scuffed areas is accelerated and their colour turns brown to black.

Management

- Careful handling to minimize the abrasion injury.
- Avoid the fruits under refrigeration below 90% relative humidity.

m) Leaf Sheath Loosening

Leaf sheath loosening is a physiological disorder in Robusta banana. Two or three leaves from the base of the pseudostem become yellow or light green and get detached. Lack of potassium coupled with accelerated senescence of leaves because of high levels of nitrogen leads to this problem. Normally this disorder occurs in 5-9 month - old plants and reduces yield only marginally.

Management

It can be controlled by applying recommended dose of NPK per plant at third, fifth and seventh month after planting.

n) November Dump

Cold temperatures also have a marked effect on bunch development. When the flower initiation occurs during the winter, a characteristically peculiar bunch can be expected. The November dump bunch is small and composed of hands of uneven size and often deformed fruits. The smaller fingers on the hands are, in many cases, fuller at the flower end and thinner at the stalk end. The flower end is often black, rough and cracked and during a good rainy season, fungal diseases further damage these fruits. Another common symptom is the cone-shaped protuberance at the flower end. This protuberance is pale green in colour. It must be stressed that November dump symptoms are caused by cold temperatures and they are not mineral deficiency or drought symptoms.

Management

- A good orchard practice is to cut down all the plants producing severe November dumps.

o) Peel Browning

Peel browning is a physiological disorder occurring during maturation process, when the fruit is handled. Some varieties seem to be more sensitive than others. Peel browning is caused by a stress at low temperature or relative humidity. Peel browning can be assessed by storing ripened bananas in a room at ambient temperature (20°C) and low relative humidity (50%). Browning limits shelf-life of the banana and reduces it's commercial value. Thus, browning is easily detectable by sellers and consumers but is not attractive, curbing act of purchase.

Bananas peel browning is reflected in a brown discoloration of the peel triggered by a peel handling. Browning intensity increases day after day when bananas are stored in dry conditions. Symptoms develop the entire banana surface. These disorders result from an enzymatic and non-enzymatic oxidation of phenolic compounds. It involves the activities of poly phenol-oxidase (PPO), phenylalanine ammonialyase (PAL) and peroxidase (POD).

Management

Storing fruit at high relative humidity (95%) can delay peel browning. In practice, atmospheric relative humidity during shipping and maturation could be managed by a maturation chamber ventilation, air extraction system and use of macro perforated polyethylene bags.

p) Peel Bruising

Banana bruise damages appear on the market stalls during the handling of fruits. Peel bruising has been identified as a major cause of quality loss, resulting in a decreasing commercial value. Bruise susceptibility is not exactly a physiological disorder since that it results from a mechanical damage more especially an impact which could be paired with a skin rupture. Bruising leads to the appearance of a brown to black impact on the fruit surface. That occurs rapidly after the damage and spreads on all the impact area. Symptoms are observable on the banana peel but do not impact necessary the fruit pulp. Browning linked to bruising would be triggered by enzymatic reaction. Susceptibility is also impacted by the fruit turgidity and flesh firmness at time of injury. The severity of bruising depends on the energy and number of impact, the type of impact surface. Bruising susceptibility relies on several factors as temperature, fruit size and fruit shape or tissue and cellular factors as cell wall strength and elasticity, cell shape and internal structure.

Management

Reducing temperature during post-harvest chain could reduce damages caused by bruising.

q) Peel Cracking

Genetic, environmental, morphological and physiological factors affect peel cracking. Peel splitting or cracking is mainly attributable to the environmental conditions of storage. It is one of the main physiological disorders limiting post-harvest fruit quality and quantity. Peel splitting results in open wounds promoting rapid moisture loss and opportunistic pathogens infection. Moreover, cracking fruit is not attractive to consumer and also decreases it's marketability.

Peel cracking is a physiological disorder reflected by a longitudinal split of the peel and exposing the pulp. Generally, it begins near the pedicel and more especially at peel junctions between two flat segments. Currently, this disorder is attributed to a peel water content reduction by an inverse water flux from the peel to the flesh. This water movement is induced by osmosis due to an increasing sugar content in the pulp. This provokes a turgor pressure increase higher than the pressure corresponding to the expansion capability of the fruit cuticle. Peel cracking is associated to a boost of respiration, an increase of the oxidative stress markers and a high oxidative stress damage.

Management

Peel cracking can be managed by implementation of dry humidity conditions during ripening, i.e. by managing the level of perforation of polyethylene bag or the humidity in the ripening and storage chambers.

r) Green Life Shortening

Banana green life duration corresponds to the potential storage life of banana before the initiation of the climacteric rise. The knowledge of this value is crucial for the export banana, because this represents the ability of the fruit transport. Therefore, it is inacceptable to select varieties with very short green life duration, except for the local market, since fruits must withstand shipping transport during three or four weeks. Bananas are climacteric fruits i.e., they are able to ripen even after the harvesting. Their ripening process is divided into three parts: a pre climacteric stage, the climacteric rise (corresponding to an increasing respiration) and fruit maturation (characterizing by an ethylene production).

s) Water Soak

Bananas growing in badly drained or water-logged soils, especially heavy clays, show characteristic external symptoms. Both young and older plants are subnormal in height for their age; they may, indeed, be quite stunted. The leaves have a flagging, wilted appearance and sometimes break half-way along the blade; they are of a pale-green colour, fading to yellow, sometimes with streaks along the veins. 'Water-soak' plants may sometimes resemble Panama Disease externally but not at all internally. The internal symptoms of 'water-soak' are somewhat indefinite: they include a water-soaked discolouration of the rhizome tissues, from which, however, no pathogenic fungus or bacterium has been consistently isolated. The vascular strands in the central tissues of the rhizome may have a darkish or blackish discoloration. At a late stage, the rhizome tissues break down and are invaded by bacteria and presumably by other organisms and culminating in a spongy rot.

In relation to the waterlogged condition, or to inadequate drainage, with poor aeration and oxidation, the soil becomes sour. This interferes with the useful activities of soil bacteria and with normal root development and functioning. In typical stunted, 'water-soak' plants, the roots are feebly developed; they extend but a short distance from the rhizome and a considerable proportion of them may be black and rotted. These defective root systems are often seen in bananas planted in areas that have only recently been reclaimed from swamp forest conditions by the installation of drainage, i.e. where planting has been done before the soil has had time to become aerated. 'Water-soak' may also occur as a result of localized water-logging in compact, water-retentive soils, after heavy rains, even though drains have been dug. These soils eventually drain in dry periods, but the duration of poor aeration may be such that root growth is restricted and decay soon sets in.

Management

These include the installation of suitable drainage as a first step. Sufficient time must be allowed for the drains to work and for the soil to become fully aerated and certain of its components weathered and oxidized. The digging of considerably larger planting holes and leaving them and the excavated soil to weather for several weeks before planting should be included in the plans for opening areas where there is a problem of excessive soil wetness. Heavy applications of quicklime, or lime, cultivation and the incorporation of cover-crops or of other humus to improve soil structure, are likely to be helpful.

t) Drought

There are also areas where bananas are grown without irrigation and where drought conditions supervene as the annual dry season advances. The plants gradually stop growing and come to a complete standstill. The leaves show evidence of desiccation and become yellow and wilted. The bunches are small and of poor quality and mature slowly over an unusually long period. In a severe drought, the young bunches may fall to the ground and break easily.

u) Salting

'Salting' a serious condition in irrigated lands and one which may result in large areas going out of cultivation. As 'salting up' became accentuated, banana cultivation gradually receded and disappeared and typical halophytes and leathery-leaved and prickly xerophytic plants became established. A dead brown band eventually extends along the entire leaf margins, as if chemical scorching had taken place.

v) Defective Planting Material

Badly selected, though not actually diseased, planting material may give a disappointing, uneven stand of plants, call for further expenditure on replacements and delayed production. The kind of plant developed and its initial rate of growth depend to a large extent on the nutritional status of the corm, sucker, 'bit' or 'eye' used and of course, on the fertility of the soil. Well-nourished 'bits', corms or suckers, with ample food reserves which will enable the young plant to make a good start, should be obtained from good localities. Small 'bulbs' and 'bits', spindly stems, broad-leaved 'water-suckers', 'onion' bulbs and in general, material taken from very young or from old debilitated stools, or from poor localities, or from localities with an undesirable disease record, should be avoided.

w) Wind Damage

A 'blow-down' not only causes the immediate loss of bunches at all stages of development - some of the harvestable bunches may be salvaged but followers of various ages may also be destroyed or damaged and various aberrant developments may continue to take place long after the main destructive effects have disappeared. Wind damages are:

- bent, curved or arched trunks;
- trunks which have sagged more or less considerably downwards from their original near-erect position, with consequential wrenching and breaking of the roots and even of the rhizome on one side and the compression or buckling of the roots on the other; and
- variously prop-damaged and scarred bunches. These disruptive effects of bad propping on the roots and on the vascular system of the true stem occur at a time of critical importance, namely, when the bunch is in the course of filling out to harvesting maturity. Some of the symptoms observed in bent, sagging, or partly uprooted stems are reminiscent of the shock effects induced in bearing trunks by the cutting out of large suckers.

x) Sunscald or Scorch

'Sunscald', as the name implies, is caused by excessive exposure of the bunch to direct sunlight. The distal ends of fingers are most affected. They appear dry and have a premature pale yellowish-green colour. In more extreme cases, species of *Fusarium* may become associated with the sun burnt tissue and a mild form of rotting ensues. In severe cases, the epidermis may be blistered. It is then replaced by a thin corky layer which fissures as the fruit enlarges. Bunches bagged in transparent materials are liable to sunscald.

y) Alkali Chlorosis

In some plantations, the bananas in localized areas, known to be alkaline, were of a bleached, whitish appearance. The symptoms of alkali chlorosis are as follows:

- Leaves, stems and bunches may show every stage, chlorophyll deficiency from a slight bleaching of the normal green colour to an almost complete yellowish white. Leaves may be affected at all stages. In some instances, the partial absence of chlorophyll gives the leaves a 'shot' or striped appearance, similar to the markings in a virus disease

- A brown scorching or rotting of young and old leaves, along the margins, on the midrib, or over the leaf-blade, frequently accompanies the development of the chlorotic condition.
- When young plants are affected, characteristically wrinkled, arrow-or irregularly shaped leaves are of common occurrence in conjunction with the prevailing chlorosis.
- In adult plants, the leaves tend to have a characteristic rigidity and erectness. Such leaves show wilting of the blades at an early hour in the morning.
- Adult plants do not attain to their normal height but remain stunted, with short, thick trunks.
- Bunches are typically short-fingered. As growth is slow, the young bunch tends to remain for an unduly prolonged period in a horizontal or semi-horizontal position, with the result that the fingers acquire a twisted appearance and the normal symmetry and compactness of the bunch are lost. Even in well advanced bunches, the abnormal curvature of the main stalk may persist; whereas fingers tend to be short over the whole length of the bunch.
- The intermediate flowers below the fruiting hands tend to remain adherent longer than is usual in normal bunches; and the formation of a small secondary bunch, between the bud and the main bunch, is of common occurrence. The male flowers also show a tendency to remain attached.

Management

Leaching with ample irrigation water in conjunction with good drainage, incorporation of abundant organic matter, applications of sulphur or gypsum, of iron salts and of fertilizers, which tend to leave an acid reaction.

z) Plant Failure

The malady, which is of great economic importance, are characterized by an early decline in the stands of bananas and by the development of some remarkable symptoms. The following factors are responsible for plant failure:

- Prevailing low relative humidity of the atmosphere and high day temperatures increases evaporation from the leaves.
- Insufficient rainfall.
- Presence of salt.
- Presence of high pH values (7.5 to 9.2) and the existence, in all areas where this type of plant failure is observed, of undesirably high concentrations of alkali, probably bicarbonates of the sodium-calcium complex.
- Insufficient quantities of nitrogen, potassium and phosphorus and of organic matter.
- An abnormal condition of leaf failure, premature death became apparent. This condition is particularly noticeable about the time of shooting of the bunch.
- In five to ten months old plants, the older (lower) leaves lost their normal full green colour prematurely and became yellow, yellow-brown dirty yellow and streaked, and spotted by secondary leaf fungi. Later, the leaves became brown and if the petiole is unbroken, permitting of injection with water from below, of a dark brown colour, constituting what is called as 'black leaf'. Some young plants were, thus seen to be carrying two or three brown leaves and one or two yellow- brown leaves, or there is a cluster of brown leaves hanging down. In older plants, yellow and discoloured greenish-yellow, brown and fallen leaves developed prematurely in acropetal succession, so that bearing plants are left with an abnormally small number of leaves to cater for the important period of bunch development. In some plants, affected leaves are heavily spotted with secondary fungi; in others, brown bands and browning of the leaf-tips, as in weevil-borer infestations are of common occurrence. Marginal browning or 'burning' of the type that accompanies 'salt scorch' is very common.
- In some areas, under the prevailing conditions of low relative humidity and exposure to the fierce sunlight, leaves are discoloured, particularly along the midrib and on the blade on the southern exposure, to a dingy, yellowish colour. Such leaves

have a dry appearance and are more or less crinkled, with abnormal twisting at the distal end. Some, young rolled leaves also showed abnormal twisting and evidence of sun scorching.

- In the second or third year, leaves of affected adult plants appeared to be lacking in length and breadth and tend to form an abnormal erect cluster round the 'throat', instead of the graceful spread, typical of normal healthy plants. The leaf-blades drooped early in the morning and this effect, in conjunction with the leaf colours described above, gave the plants what has been described by cultivators as a 'sleepy' appearance.
- Where the petiole of affected leaves joins the trunk or pseudostem, a black mark is typically present. With the progress of ageing or senescence in the leaf, this black mark increased in size and is seen to be associated with a localized water-logging of the petiole-pseudostem tissue, this being caused, or accompanied by the activity of micro-organisms of the soft-rotting fermentation type. As these localized blackened regions increased in number as leaf after leaf became affected, the trunk acquired an unhealthy, pale yellow or pallid green appearance. As bunches are frequently slow in filling out, the main stalk took longer to reach the vertical position and fingers accordingly developed curvatures which marred the symmetry of the bunch. At harvesting stage, bunches are short-fingered, light in weight and irregular symmetry.
- In some instances, the stature of the plant is also reduced.
- More or less complete destruction of leaves, rotting and softening of the leaf-sheaths at the 'throat' or 'crown' of the plant and the dropping out of bunches are observed. Both small and large bunches are liable to drop out, the main stalk of the bunch typically breaking, with some localized soft-rotting at the whitish region. Within the throat, such breaking is primarily mechanical: it is a direct result of the failure of the softening petiole-bases to give their normal support, the main stalk usually breaking under wind action at the weakest point in proximity to the fulcrum of the main-stalk bunch unit. Severely affected plants shows a tendency to develop a leaning position, even when the bunches are quite small and light.

❒❒❒

5

Pomegranate

a) Fruit Cracking

Cracking of pomegranate fruits (25-40%) takes place during maturity. The cracked fruits though sweeter in taste, loose keeping quality and become unfit for shipment and liable to rotting. The cracking in fruits may be due to sudden change in soil moisture content. At the time of fruit ripening, if soil becomes too dry and then irrigation is given heavily, cracking occurs. Cracking is due to the following factors:

- Heavy irrigation or rain followed by an extremely dry spell results in cracking.
- Rind thickness and texture of fruits in various cultivars influence the intensity of cracking.
- Very heavy incidence of mites on one side of fruits causing imbalance which leads to fruit cracking.
- Deficiencies of Ca, B and K also cause cracking.
- Cracking in young / immature fruits could be due to boron deficiency, while in fully-grown fruits, it could be due to moisture imbalance because of variation in soil moisture and atmospheric moisture deficit.
- Varietal characters are also responsible for cracking of fruits at different stages of growth.
- Prolonged drought condition cause hardening of peel and if followed by heavy irrigation of rainfall, then pulp grows resulting in cracking.

The increase in air temperature also creates fruit cracking. It has also been related to deficiency of calcium, boron and potassium besides difference among cultivars. The extent of cracking in different cultivars varies. Some of the cultivars are less susceptible to fruit cracking. Rind texture and thickness in various cultivars seem to influence the intensity of cracking.

Management

- Adequate and regular light irrigation throughout the bearing period may reduce the intensity of cracking and improving of water holding capacity of the soil by adding organic manure.
- Spraying of Giberellic Acid @ 120-250 ppm or boric acid @ 0.1% is found to reduce the fruit cracking. In some places, harvesting is done prematurely to avoid cracking but the fruits are poor in quality.

b) Regulation of Flowering

Pomegranate flowers continuously when watered regularly. The plants under such condition may continue bearing flowers and bear small crop irregularly at different period of the year, which may not be desirable commercially. To avoid this, trees are given **bahar** treatment. In this treatment, irrigation is withheld two months prior to the bahar followed by light earthing up in the basin. This facilitates the shedding of leaves. The trees are then medium pruned 40-45 days after withholding irrigation. The recommended doses of fertilizers are applied immediately after pruning and irrigation is resumed. This leads to profuse flowering and fruiting. The fruits are ready for harvest 4-5 months after flowering.

In tropical condition, there are three flowering seasons, viz., January-February *(ambia bahar)*, June-July *(mrig bahar)* and September-October *(hasta bahar)*. The choice of flowering/fruiting is regulated taking into consideration the availability of irrigation water, market demand and pest/disease incidence in a given locality.

Type of flowering (bahar period)	Time of flowering	Time of harvesting
Ambia Bahar	January-February	June-August
Mrig Bahar	June-July	November-January
Hasta Bahar	September- October	February-April

The fruits of **ambia bahar** are ready for harvest in the month of June to September. As the fruit development takes place during dry months, they develop an attractive colour and quality, thus suitable for exports. Similarly due to dry weather, the incidences of pest and disease attack are limited. However, ambia bahar can be taken only from areas having assured irrigation facilities.

The **mrig bahar** crop is harvested in the month of November to February. Usually this bahar is favoured as the flowering and fruiting period coincides with rainy season or immediately after rains and the crop is taken without much irrigation. As the fruits develop during the rainy season and mature during winter, the colour and sweetness of the fruit is affected.

The fruits from **hasta bahar** are harvested during the month of February to April. They have very attractive rind with dark coloured arils. Since the availability of the fruits during this season is limited, they fetch high value.

c) Chilling Injury

The incidence and severity of chilling injury depends upon storage temperature and duration. The minimum safe temperature is 5°C (41°F) for upto 8 weeks. Chilling injury can be a major cause of deterioration of pomegranates during marketing following exposure to temperatures below 5°C (41°F) during storage and transport for longer than 4 weeks. External symptoms include brown discolouration (scald) of the skin, pitting and increased susceptibility to decay. Internal symptoms include brown discolouration of the white segments separating the arils and pale colour (loss of red colour) of the arils. Pomegranate are susceptible to chilling injury if stored longer than one month at temperatures between their freezing point (-3°C or 26.6°F) and 5°C (41°F). Upon transfer to higher temperatures, respiration and ethylene production rates increase and other chilling injury symptoms appear; their severity increases with lower temperatures and longer durations of chilling exposure. External and internal browning is related to oxidation of phenolics by polypherol oxidase. Storage in a 2% oxygen atmosphere at temperatures below 5°C (41°F) reduces severity of chilling injury symptoms.

Management

Avoid exposure of pomegranates to temperatures below 5°C (41°F).

d) Aril Browning

Recently, the incidence of a physiological disorder called 'aril browning' or 'aril blackening' (internal breakdown of arils) has threatened the cultivation of pomegranate. The affected arils are soft, light creamy-brown to dark blackish-brown, deformed, acidic and possess unacceptable off-flavour and unfit for consumption. The damaged arils vary from a few to all in a fruit. This disorder does not show any external symptoms and often possess good eye appeal. Defective arils are detected only after cutting the fruits open, posing a serious challenge to quality control in export to foreign countries. Sometimes, the local consumers hesitate to buy this fruit because of the hidden brown arils. The study conducted by IIHR, Bangalore, confirmed that the role of seed in initiating the disorder.

Management

Preharvest application of plant growth regulators reduces the incidence of aril browning.

❑❑❑

6

Grapes

a) Bud Dormancy

Dormancy may be defined as any phase in the life cycle of a plant, in which active growth is temporarily suspended. The resting buds are formed mostly in trees of temperate regions. Grapevines are grown mainly in temperate regions, which have cool to cold winters and warm and relative dry summer days. In South India, bud break for vegetative growth take place during April-May. After the brake of the dormancy, the buds enter into breaking, bursting and finally produce flowers and fruits. By hastening the process of bud break, it is possible to get two crops per year and thereby, increase the yield.

Management

Various management practices followed to overcome this problem of bud dormancy in grapes are as follows :

- Day length has a marked effect in bringing about dormancy. It appears that some role is played by phytochrome in the initiation of dormancy. Hence, long day treatment was found to be effective at maintaining growth and inhibiting the onset of bud dormancy; whereas short day treatments promoted the onset of bud dormancy.
- Buds require a period of winter chilling before its growth. The effective temperature for overcoming dormancy ranges from 1-10°C and the length of required chilling period ranges from 260 to 1000 hours.
- Several substances, such as ethylene, chlorohydrin, thiourea and dinitrophenol can overcome dormancy. Thiourea (2%) or KNO_3(2%) is highly effective in increasing the percentage of bud burst in grapes.

b) Excess Canopy

Over vegetativeness is a major obstacle in grapevines. Cropping in perennial plants is a more complicated process than in annuals, which are preprogrammed to initially develop a simple photosynthetic framework. In perennials, there is a continuing requirement to divert part of the total growth into maintaining and developing the permanent architecture of the plant. There is a significant competition between vegetative growth and fruit growth. It is, therefore, necessary to strike a balance so that at least a portion of growth and photosynthesis is available to maintain and expand the basic structure of the grapevine.

Management

The only attempt made to overcome the problem of over vegetativeness is pruning. One of the main reasons for pruning grapevine is to manipulate their shape for easy management. Vine pruning regulates good yield and improves the size and quality of fruits. Pruning seeks way to increase the efficiency with which grapevines convert the solar energy to chemical energy. Apart from this, photosynthates are being properly partitioned to maintain the correct balance only after pruning.

c) Fruit Bud Differentiation

Fruit bud formation takes place for next year on current season growth in temperate conditions and 45 to 60 days after pruning on current season growth in peninsular India. Bud fruitfulness varies with inherent capacity of variety. Studies on fruiting habits of different varieties have shown that first two or three buds from the base contain fewer inflorescence primordia than upper bud. The proportion of fruitful bud increases progressively towards the middle of cane and falls off towards the tip. Shoot vigour, light and temperature plays a major role in fruit bud differentiation. More the vigour of shoot with respect to its internodal length and weight, lesser will be the productivity.

Management

(i) Use of growth retardants

Use of recommended dose of nitrogen and required quantity of water, spraying growth retardants like CCC after back/foundation pruning (March-April) can increase fruitfulness. These sprays increase bud fruitfulness through various physiological changes viz., reducing vegetative growth, increasing shoot thickness, reducing internodal length by counteracting the effect of endogenous 'GA' and increasing endogenous cytokinin levels and consequently ratio of cytokinin/ gibberellin congenial for formation of inflorescence primordia.

(ii) Increasing light penetration in vine canopy

Orientation of growing shoot should be in such a way that the buds on them should get maximum available sunlight.

(iii) Promotion of cytokinin/GA ratio

Application of Benzyl adenine to shoot apices will prevent the formation of tendril primordia and favours the formation of inflorescence, since cytokinin plays a vital role in controlling the process of inflorescence formation, differentiation of flowers and pistil development.

(iv) Nutrient management

Nutrition has been considered as the major factor in determining the productivity in vines. Appropriate quantity of nitrogen induces the flower initiation through synthesis of proteins and nucleic acids which are favourable for inflorescence formation. combination of N, with P and K will induce early flower bud initiation.

d) Flower Bud Drop

One of the major constraints leading to poor yield in grape is the flower bud drop. The reasons for this may be as follows:

- Abscission of flower bud pedicels possess a great problem in vigorously growing cultivar namely Thompson seedless and Pusa seedless.
- A temporary water deficit in developing flower clusters that lead to higher rate of transpiration from developing leaves appears to be the cause for this malady.
- High Nitrogen supply under high temperature leads to rapid elongation of shoot which competes with developing flower buds for water supply.

Management

- Decreasing the rapidity of shoot growth by limiting nitrogen and water supply during spring.
- Topping vigorous shoots.
- Growing of cover crop to compete with the vigorous shoot growth.
- Irrigate the vines regularly to avoid water deficit.
- Drop of flowers or buds follows the formation of abscission layer. Hence, application of NAA @ 20 ppm retained maximum number of buds and flowers.
- Spraying Planofix @ 300 ppm after fruitset decreased the berry drop but increased the ascorbic acid content in grapes.
- Application of GA_3 @ 20 to 50 ppm, 18 days before anthesis decreased flower drop in Thompson seedless grapes.
- Girdling at 10 days before full bloom will improve the fruit set in Thompson seedless grapes.

e) Bareness of Vine

Commercial grape cultivar like Anab-e-Shahi in South India become erratic in bearing after giving good yields for few years and resulting in poor yields in the next few years. Failure of flower bud differentiation and death of floral primordial are the reasons for bareness in grapes.

Management

To control the bareness of vine, the remedial measure, such as discouraging excessive vegetative growth, applying balanced nutrients, checking sprouting of buds during and after rainy season, application of chloromequat, an inhibitor to biosynthesis of endogenous GA, to promote flower formation.

f) Water Berry

The water berry is characterized by flaccid, wrinkled, withered, soft, lack sugar, flavour and storage quality. Such berries in coloured cultivars are called "red berries" due to poor colour development.

This situation is created due to the stem necrosis, which results from the interrupted flow of sugars and other ripening constituents into the berries. The earliest symptoms of water berry or cluster stem necrosis appears during onset of ripening, when the osmotic value of the berries exceeds that of the grape stalk. The first visual symptom of water berry is the development of small (1-2 mm) dark spots on the pedicels and/or other parts of the cluster framework. The symptoms may first develop in the primary, secondary or tertiary portions of the cluster stem structure, depending on cultivar and severity but usually the cluster tips, shoulders and upper laterals are most affected.

Water berry disorder causes not only a sharp decrease in the yield and quality, but it also brings fundamental changes in the physic-chemical composition of the berries. Shrivelled berries are considerably lighter with less juice than healthy berries. Also, the affected berries possess higher acidity and lower amounts of sugars and anthocyanins as compared to normal berries even with delay in ripening. The affected berries had higher acid and lower sugar content than healthy berries at ripening. The amount of total phenolic compounds, anthocyanin content of red cultivars and level of total fatty acids are lower in affected berries than healthy berries.

Management

- $MgSO_4$ sprays at the beginning of berry ripening and repeated 7-10 days later reduced the disorder by 70%.
- Heavy crop load causes formation of water berry by developing photosynthetic stress in the vine. Judicious pruning (removal of weak and excessively vigorous shoots) and long tipping (10 internodes above the first grape cluster) will reduce this disorder.
- Post bloom cool weather reduced the cell division and calcium uptake in the berries which lead to the development of water berries.
- There are several preventive measures to check this malady in different cultivars. These include the selection of suitable cultivar, judicious fertilizer and manure application, spraying or treating the bunches with nutrients, fungicides, antioxidants, antitranspirants or growth regulators, moderate pruning, removing weak and excessively vigorous shoots and sowing of cover crops.

g) Rudimentary Panicles

Formation of rudimentary panicles are due to inadequate nourishment for the development of the panicle because of competition between too many sinks. The more damage has been observed because of excess nitrogen. Excess nitrogen reduces the carbohydrate accumulation. Shoots tend to have long internodes and become flat and cause excessive shoot growth and delays cane maturity. Panicles after their emergence fail to grow to their normal size. Their growth is stopped at about one cm length. They do not turn green development of floral parts is confined to calyx only. Corolla, stamens and pistil do not form. These rudimentary panicles are dropped down. Shoots bearing rudimentary panicles are weak and short. Leaves are pale-green in colour and small in size. The disorder is observed due to following factors :

- Too many fruiting canes, particularly in young (one or two years old) vines.
- Too many growing shoots and panicles on a cane, particularly on a thin cane.
- Water stagnation in the root zone for a week immediately after bud break.
- Prevalence of chill weather at flower emergence.
- Presence of excess nitrogen in soil promotes vegetative growth and adversely affects the fruitfulness.

Management

This problem may occur due to cultural practices, development of excess of number of fruiting canes and forcing excess number of buds to open after pruning with an aim to obtain high yields even in young vines. Therefore, limiting the number of canes to 5 per m^2 in Thompson Seedless and allowing only two shoots to grow on each cane (one shoot only on canes of less than 6 mm thickness) either by limiting the bud break or shoot thinning at 3-leaf stage can avoid this problem.

h) Shot Berries

Berries that are smaller, round and seedless as compared to the normal berries in a cluster are commonly called short berries. Boron deficiency and incorrect stages of gibberellic acid (GA) application and girdling are the known reasons for shot berry formation. It has been proved that flowers may fail to develop into normal berries because of poor pollination or fertilization, defective flower parts, or poor carbohydrate nutrition to the flowers. Shot berries, in a seeded variety are seedless and small, while in a seedless variety, they are smaller in size than the normal ones. In any case, they are sweeter than the normal berries. Often berries, that might drop, remain on the cluster and develop into shot berries. Normally, the small berries shatter after 7-10 days after set due to failure of embryo to form or lack of nourishment. But some berries do not drop and even they do not have embryos, get adequate nourishment and eventually develop into shot berries.

Management

- Fertilization of ovary is impaired due to restricted growth of the pollen tube. Application of auxins at bloom helps the growth of even unfertilized ovaries by the stimulus they impart.
- Application of GA_3 or girdling immediately after set and before the shatter stage retains those berries which have fallen due to lack of nourishment.

i) Berry Cracking

This disorder is basically associated with rains during ripening of grapes. It is very common in Thompson Seedless. If rains occur during the late ripening period, berries gets cracks and then rots. In South India, Bangalore Blue is less prone to this disorder even if rains occur during their ripening period. Factors affecting this problem are:

- Advanced stage of ripening with TSS content more than 16° Brix,
- Adherence of skin to pulp and
- Heavy rains (more than 10 mm a day) associated with the build-up of atmospheric humidity.

The impact of splashing rain drops on the soft berries is supposed to be the cause for berry cracking. But the turgor pressure build-up in the ripe berries as a result of water absorption by the roots and lack of transpiration from the leaf surface due to rain and high atmospheric humidity seems to be the basic reason for berry cracking. High soil moisture and high atmospheric humidity are basic pre-requisites for berry cracking and rotting.

Berry cracking is also observed in clusters protected by the foliage from getting hit directly by the rain drops. Rotting of berries is also observed in compact bunches of Thompson Seedless which ripe during February due to early pruning. The rotting symptoms appear on the third day after a heavy rain of more than 10 mm within a day.

Management

Early pruning to avoid the coincidence of ripening with the normal rainfall occurring period in the peninsular India and hastening of ripening by suitable means in north India can reduce the chances of berry cracking and rotting.

j) Pink Berries

This disorder is commonly observed in Thomson Seedless and its clones mainly in Maharashtra. The berries become pink during berry ripening and close to harvest. The incidence of pink berry development seems to be more in vineyards, in which large diurnal variation in temperature occurred during the period of veraison stage, if the day temperature ranges from 30-35°C while the night temperatures down below 10°C. Hence, the pink berry formation was relatively more in vineyards in which berry softening took place in the first week of February (low temperature conditions (< 10°C) prevails during this time) and was less in vineyards in which the softening occurred in the first week of March (temperature starts rising).

The berries become pink in colour. Initially red bunches appear on the skin on a few berries in a bunch. Gradually, the pink blush spreads to the entire surface of berry and many berries in a bunch become pink. The pink colour is minimum and starts at veraison stage and reaches to its maximum within a fortnight. The pink colour changes to dull red colour rendering bunch unattractive.

Management

The vines should be pruned either in the last week of October or the first week of November. Soil moisture stress from berry softening until 10 days prior to harvest should be avoided.

k) Berry Shrivel

It begins after ripening. Red Globe berries become flaccid and sunken. Some affected berries develop colour; some remain white. The flaccid berries are usually interspersed with normal ones, but occasionally several berries at the tip of a lateral or at the cluster apex are affected. The clusters may be a slightly dull green but the berries must be touched to confirm the flaccid condition. The amount of fruit involved usually increases up to harvest.

Management

GA_3 applied about two weeks after fruit set, when the berry diameters average 10-15 mm will reduce the shrivel by 50 to 70%. The bunches covered with papers may also reduce the intensity of berry shrivel. This should be done 80-90 days after pruning.

l) Sun Scars / Scorching on Berry

Sun scars or sun scorching may be the most damaging disorder amongst all disorders, which burns the berries and bunches before veraison and causes considerable losses in the grape industry. Due to inadequate foliage and improper orientation of the shoots or even in the presence of adequate foliage, sometimes the clusters get exposed to sun rays and it causes physiological disorder in berries. There should be at least 10 leaves above the cluster and the shoot bearing cluster should be oriented horizontally or diagonally at 40-45 degrees angle. If the bearing shoots are allowed to grow upright on a vertical trellis or straight down on a single wire trellis, even with more leaves on the shoot, the bunches are exposed to sun. Therefore, in addition to adequate foliage, shoot orientation (shoot positioning) is important. Shoots have to be tied to the trellis in such a way that the bunches are not exposed to afternoon sun. This operation has to be carried out as the veraison (fruit softening/colouration) starts.

When the bunches are exposed to sun during ripening, berries get tanned and develop amber brown colour or red to yellow colour on the portion, particularly on the south-west side of the bunch exposed to sun, while berries will be green on the unexposed portions of the bunch. The symptom of scorching by thrips shows irregular corky scars while that of powdery mildew shows blemishes on the berries and also cause scars on berries.

Management

Adequate plant protection measures are to be taken to avoid the incidence of sun scars and scorching by thrips and powdery mildew on berries and adequate care and caution are to be exercised while spraying fungicides to control the diseases taking into account the stage of berry development, the choice of fungicide and its concentration. Some control measures are given below:

- The canopy management should be taken into consideration. There should be atleast 10-12 leaves maintained above the cluster and bunch.
- The orientation of arm should be North South direction to avoid direct exposure to sunlight.
- Avoid spraying of Bordeaux mixture and fixed copper formulation, or wettable sulphur during the period commencing berry set to pea stage of berries to avoid formation of black specks of tender berries.
- Spray wettable sulphur only after pea stage of berries cautiously only at recommended concentrations, if needed.
- Avoid drenching of the bunches with spray solutions, particularly of wettable sulphur or copper fungicides.
- Do not allow the fungicides to settle in the tank or container while spraying to avoid chemical scorching of berries sprayed with the solution at bottom.

Uniform and quick bud burst

Uniform and quick bud burst is a problem after forward pruning due to severe winter in some regions. Swabbing of buds with Hydrogen Cyanamide (@ 1.5 ml/litre) within 24-48 hours after pruning is recommended. Similarly, to reduce the apical dominance (shoot vigour), Spraying of CCC @ 500-1000 ppm after 25-35 days after back pruning or vegetative pruning was recommended to improve bud fruitfulness in canes. Similarly, spraying of 50 ppm Uracil at 45 days after back pruning is recommended to improve the bud fruitfulness.

m) Interveinal Chlorosis

This disorder is common in Anab-e-shahi variety. This disorder is pronounced more at fruit maturity stage, attributed to the deficiency of magnesium.

Management

Two to three sprays of 0.5% magnesium sulphate during the fruit development stage at 20-30 days interval is recommended to reduce the symptoms.

n) Blossom end Rot

This is also common in Anab-e-shahi grape variety. A black sunken spot develops at the blossom end on the berry and this spot goes on spreading with the water soaked region leading to rot. This has been attributed to defective calcium assimilation by the growing berries.

Management

Some control measures are given below:

- Irrigating with good quality water.
- Avoiding copious irrigation after prolonged drought.
- Applying lime to the soil.

o) Dead Arm

Commonly occurs during summer, particularly in Anab-e-shahi in the States of Andhra Pradesh and Karnataka. A portion of primary arm/secondary branch suddenly dries up and eventually dies. Secondary branches are mostly affected. Roots and stem of the wilted plant show a partial discolouration in distinct sectors through the roots and affected branches. It was severe in the vines suffering from soil moisture stress and in vines growing on loose and shallow soils. Extremely high temperature during summers and high chloride contents of irrigation waters are found to aggravate the disorder.

Management

Removing the affected branched is the only remedial measure for this disorder.

p) Uneven Ripening

It occurs commonly in Bangalore blue, Bangalore purple, Beauty seedless and Gulabi cultivars. It is a varietal character with ripened and unripened fruits in a single bunch. It is associated with inadequate translocation of carbohydrates and impaired formation of anthocyanin pigment.

Management

Spraying 250 ppm ethrel (colour break stage) reduces the uneven ripening.

q) 2, 4-D Injury

Aerial applications to field crops have injured grapevines several miles from the point of application. Grapevines will exhibit damage symptoms when exposed to only minute amounts carried by the wind. The symptoms, of phenoxy herbicide damage are most dramatic on the youngest leaves and the tips of growing shoots, young leaves at the tips of shoots become small, narrow and misshapen and have closely packed, thick veins that lack chlorophyll than usual. Further down the shoot, damage symptoms are progressively less severe; leaves have irregularly shaped, a distinctive fan-shape appearance with parallel, strap like, clear veins and crystalline in texture. The leaves sometimes are cupped and the leaf margins often terminate in sharp points. Small, puckered, interveinal spots retain some green chlorophyll. During the active shoot growth period, phenoxy herbicide damage often causes growth to stop temporarily and to be retarded for several weeks. If the effects are not too severe, normal growth will resume in the same or following year. Damaged flower clusters set very few or no berries.

Grapevines in close proximity to a sprayed areas are at highest risk. Flower clusters are particularly sensitive; exposure during bloom can greatly reduce fruit set. Injured vines also may have delayed fruit ripening. Severe injury can prevent complete maturation of the fruit. The delayed maturation effect may exist in a vine for 1 to 3 years before normal ripening resumes and vines may not recover for 2 years or more. Slight injury may have little or no effect upon fruit maturity.

Management

- Encourage the neighbours to use herbicides with a different active ingredient.
- Off-target drift can be minimized by careful application timing

and methods. Wind speed and direction should be monitored closely and applications may be postponed until drift potential is very low. All label directions, restrictions and precautions should be read before using any pesticide. Spraying thickening agents (drift retardant) may help to reduce spray drift.

Finally, encourage a good neighbor relationship that will enable all parties involved to effectively conduct their operations.

❑❑❑

7

Pineapple

a) Sunscald

Sunscald is caused due to direct fall of sunrays on exposed area of the fruits making them unfit for consumption. The cells under the skin of exposed surface gets damaged due to which the peduncle bearing the fruit falls on the side exposed to the sun. The widely spaced plants are more prone to this disorder as compared to closely spaced plants.

Management

The fruits should be covered with dry straw or banana leaves or with its own leaves during the hot desiccating months of April-May. This malady can be reduced considerably by adopting high density planting in pineapple.

b) Multiple Crown or Fasciation

Multiple crown or fasciation, a physiological disorder of pineapple occurs due to genetical as well as soil and environment factors. In normal cases, pineapple fruit bears a single crown, however, in some cases, it bears more than one crown resulting in broad top of the fruits. Sometimes, the fruit gets flattened and fascinated. Such fruits are corky and become unfit for canning.

Management

- Restricting the growth of the plant.
- Avoiding advance rationing.

c) Black Heart

Black heart also known as internal browning, is a serious disorder of pineapple. Symptoms are water-soaked, brown areas that begin as spots in the core area and enlarge to make the entire centre brown in severe cases. Later, these brown discoloured spots turns black and cover the entire centre of the fruit. It is usually associated with the exposure of pineapples before or after harvest to chilling temperatures generally below 7°C (45°F) for one week or longer.

Management:

Waxing is found to be in reducing chilling injury symptoms. A heat treatment at 35°C (95°F) for one day is helpful to ameliorate this problem.

d) Flesh Translucency

The fruit becomes more sensitive to the mechanical injury and the fruit pulp becomes translucent resulting in senescence. This condition begins before harvest and continues after it. The maximum and minimum temperatures preceding three months before harvest contributes to this malady.

Management:

The incidence can be lowered by waxing the fruits after harvest.

e) Chilling Injury

The maximum storage life of pineapple at 7°C is about 4 weeks; however, when removed, chilling injury develops within 2-3 days. The symptoms of chilling injury include:

- Wilting, drying and discolouration of crown leaves.
- Failure of green-shelled fruit to yellow.
- Browning and dulling of yellow fruit.
- Internal flesh browning.
- Preharvest shading and pre and post-harvest low temperatures are the major factors increasing chilling injury symptom intensity.

f) Bruising

Fruit bruising is a major problem during harvest and packing. Bruising can be caused by impact damage, a 30 cm drop causing significant damage. The damaged flesh appears slightly straw coloured. Mechanical injury of translucent fruit can lead to leakage of fruit cell contents and loss of marketable fruit.

g) Sunburn

Sunburn is common during hotter periods (>35°C) of the year, when the fruit is not shaded by leaves and especially in ratoon crops. This condition is more prevalent in the outer rows and when fruit is lodged. Sun scorched fruits first show a bleached yellow white skin, which turns pale grey and brown, with damage to the flesh underneath. These damaged areas are more susceptible to disease organisms, particularly yeast and bacteria.

h) Collar of Slips

It is characterised by the presence of a large number of slips arising from stem close to the base of the fruit, or even from the fruit itself. The excessive slip growth is at the expense of the fruit growth and development resulting in very small, tapered fruits, often with knobs at the base. Application of high dose of nitrogen fertilizers, high rainfall along with low temperature are the major causative factors responsible for development of this physiological disorders.

□□□

8

Sapota

a) Wilt or Die Back

It is common physiological disorder. Due to anaerobic conditions in monsoon and post-monsoon seasons, wilt is of common appearance aggravated by *Fusarium spp*. This can be controlled by effective drainage facility before planting.

b) Misshapen Fruit

The shape of fruit is related with number of seeds in it which depends upon conditions for pollination at anthesis. High temperature and rainfall during flowering cause oblongation of fruits. Therefore, cultivation of sapota in areas with extreme summer temperature should be avoided. Sometimes, fruits do not develop into their normal shape but develop a depression or furrow towards the calyx end. This symptom usually appears immediately after heavy rainfall and is aggravated by high intensity of irrigation. Therefore, over irrigation should be avoided.

c) Sun Scorching

The fruits exposed to intense sunlight do not ripen uniformly, developing corkiness during winter. This is probably due to killing of hydrolyzing enzymes by heating of fruit surface in winter.

d) Poor Fruit Set

Sapota is sensitive to boron deficiency. The fruitset is poor, fruits are misshapened and young leaves are chlorotic. Spray of 0.2% boric acid in June and September helps to correct boron deficiency.

e) Chlorosis

If sapota is planted in highly calcareous soil, iron deficiency is a serious problem. Youngest leaves show interveinal chlorosis and meristem dies resulting in side shoots. Spray of $FeSO_4$ @ 0.5% with 0.3% urea once in 2 months will correct it. Better avoid highly calcareous soil.

❑❑❑

9

Papaya

a) Sunscald

Sunscald, a dark-olive-brown discoloration occurs on fruits developing on trees with very sparse foliage on trees that are leaning over with fruit directly exposed to the sun and harvested fruits are left exposed to the sun.

b) Blossom end Defect

It can be severe at some times of the year. It leaves an open channel at the blossom end to the fruit seed cavity. Fruits with this disorder are prone to bacterial diseases in the fruit seed cavity before harvest.

c) Hard Lumps

Hard lumps in the flesh occur in frequently in otherwise ripe, non-heat-treated fruit. The lumps are thought to be associated with insect injury, disease, or other foreign material in the flesh. This condition can also occur in heat-treated fruits. Premature ripening of fruit, referred to as "soft fruit," is related to low calcium content. This condition is more common following periods of heavy rainfall, 2 to 3 months before harvest.

d) Intra-ovarian Ovaries

Intra-ovarian ovaries are common in some strains of papaya. These ovaries occur as a proliferation of tissue in the seed cavity and can be a thread like appendage to round or elongated structures of various sizes and shapes. A few fill the entire seed cavity of the fruit and have their own seed cavity.

e) Fruit Deformity

Fruit deformity is identified as a new physiological disorder of papaya under diverse agro-climatic conditions of India. Deformity first starts in young fruits, but symptoms become more severe over fruits close to physiological ripening. The bumpiness begins in areas on the fruit epidermis and is due to boron deficiency, arresting the fruit growth. In addition, infected tissue continues increasing in size and end up forming a protuberance of "bump", similar to a ball. High incidence before maturity has also been a serious concern in papaya fruit production.

f) Catface Fruits

Many times, some misshapen fruits are produced on the plant which are unmarketable. There are many factors which may be responsible for the production of such deformed fruits. One of the reason is the improper development of carpels and stamens which are also referred as the problem of carpellody and sterility. Carpellody refers to the abnormal development of the stamens into carpel like, fleshy structures. In other words, the stamens become part of the fruit. This results in misshapen 'Catface' fruits. Often carpellodic fruits may resemble female fruits. Carpellody appears to be influenced by environment. It occurs especially in those varieties that are produced in areas with warm temperature than replanted in areas of cool temperature. Sterility refers the phenomenon in which normally productive hermaphrodite trees changes into a condition of maleness. Instead of producing hermaphrodite flowers, only male flowers are produced. Warm and dry weather appears to encourage the expression of sterility.

❑❑❑

10

Avocado

a) Excessive Fruit and/or Flower Drop

If the fruit drop becomes excessive or premature, this is probably because of fluctuating temperatures and drying winds. Inadequate moisture to the roots is also a likely cause of this disorder. Protect from the wind as far as possible and irrigate the plant, when it needs it.

b) Tree Blooms but Produces no Avocado Fruit

Inadequate pollination is the likely cause of this problem. By planting more bees-appealing plants for attraction of bees or even try raising honey bees will solve the problem.

c) Plant Produces Finger Shaped Avocados (cocktail avocados or cukes)

It is caused by sudden changes in temperature or a lot of drying winds. They are still edible and don't have a seed, so can be peeled and eaten with a dip water regularly.

d) Tips of Leaves Turn Brown

Inadequate moisture, dry winds and also excessive salt in the soil may be the cause of this problem. Leaching the salt from the soil by employing irrigation techniques, about three feet from the trunk will solve the problem.

e) Brown Spots Disorders

Dark-brown spots show up fruit peel makes severe problem in fruit marketability. The symptoms as small bright-brown spots develop at picking time, become dark-brown during storage in cold temperature of 4-5°C and could reach more than 25 per cent cover of peel area. The spots are visualized more profoundly on background of hypertrophic lenticels. Damage is correlated with lenticels intensity on peel and with fruit size. The damage severity showed up differently in different seasons, locations and picking times. The damage was affected by the clonal rootstock/scion combination. Bruising during fruit packing and handling accelerates the peel damage.

Management

Foliar application with NPK, fruit dipping in GA_3 and application of antitranspirants may reduce this problem.

f) Fizzles

Leaves will be much reduced in size and pale in colour accompanied by multiple buds and broom-like growths. More common in light gravel soils, where drainage is excessive. Trees may almost die and then recover. Large quantities of manure left as a mulch on the surface with very frequent light watering may sometimes bring marked improvement.

g) Ring Neck

High incidence of ring neck disorder was associated with high leaf K and lower Mg contents. The symptoms first start on the lower side and upper side of the leaves and in severe case, defoliation of the leaves.

Management

Application of Ammonium sulphate or urea in more quantity may reduce this disorder.

h) Pulp Spot

The region surrounding the cut vascular bundles of the fruit stalk becomes black. This is usually localized in nature and the symptom is higher early in the season.

❏❏❏

11

Custard Apple

a) Fruit Splitting

Probably caused by sudden changes in fruit moisture content or temperature and some species and varieties appear to be less susceptible than others.

b) Russeting

The combination of low night temperatures (lower than 13°C) accompanied by low humidity is the probable and main cause for superficial russeting of the fruit skin. Mature fruits are more susceptible than young fruits.

c) Crocodile Skin

Fruits show wavy and pointed carpels. Extremely vigorous plants show more severe symptoms.

d) Hard Seed casing and Brown lumps

Boron deficiency or sudden changes in fruit water content may be the main causes of this disorder. Pulp of the fruit commonly shows hard seed, causing this disorder when trees are cultivated in dry areas without irrigation.

e) Stone Fruits

With the end of harvesting season, the leaves of custard apple turn yellow and with the onset of winter, leaves fall down. New growth commences in spring after passing a dormant phase in winter. But in certain neglected plants or those under severe moisture or nutritional stress, the dormancy sets well in advance and some fruits turn brown, become quite hard and without further growth remain on tree for months even in the following season. Such fruits are referred to as 'Stone fruits'.

Management

This disorder can be corrected by practicing clean cultivation, manuring, timely irrigation and application of superphosphate and bone meal. If plants are well-attended, they flower in time and their fruits mature before the plants enter dormancy without stone formation. Sometimes, very young fruits turn black, become hard and stop further development. This mummification of fruits appears to be a natural thinning mechanism, especially during periods of drought.

f) Tree Decline

Tree decline is caused by water stagnation. Such trees shrivel and drying of old branches takes place and they die suddenly. Stagnation of water during heavy rains should be avoided.

g) Fruit Cracking

The cracking of fruits is noticed in custard apple. It is caused due to sudden and high fluctuations in water supply to the plants. Usually, cracking occur from heavy rainfall or irrigation after a prolonged dry spell. Appropriate irrigation scheduling can reduce the problem of fruit cracking. The abrupt decline of custard apple trees may result due to water logging in sandy or rocky soils.

❑❑❑

12

Aonla

a) Internal Fruit Necrosis

In this disorder, necrosis symptoms start with browning of innermost part of the mesocarpic tissues at the time of endocarp hardening in the second and third week of September. Browning of mesocarp extends towards the epicarp resulting in brownish black areas on the fruit surface in the second and third week of October. Depending upon the severity of the disorder, mesocarp of the affected fruit turns black and becomes corky and gummy pockets develop.

Management

Spray of 0.3 % boric acid thrice in the month of September and October at about 10-15 days interval is recommended to control the disorder.

b) Unfruitfulness in Aonla

This problem is more pronounced in the variety, 'Banarasi'. This problem is not due to lack of pollination as often believed, but it is largely due to the absence of female flowers in the tree. In aonla, fruits are borne in leafy shoots. These shoots bear male flowers in their upper end and female flowers in the lower end. On the basis of male and female flowers, these bearing shoots can be divided into two types:

- Shoots bearing both male as well as female flowers and
- Shoots bearing only male flowers. In the Banarasi variety of aonla, the number of female flowers or shoots containing female flowers are inadequate and, as such, it is highly prone to unfruitfulness.

Management

- While selecting a branch as scion, it should be ensured that the scion is taken from the branch bearing maximum number of female flowers.
- An aonla orchard should have trees of two or three varieties in order to avoid unfruitfulness of a variety due to lack of female flowers.

❑❑❑

13

Bael

a) Fruit Drop

Fruit drop is a very serious problem in bael. About 99 per cent fruits drop down. Fruit drop starts from July and continues till harvest. The maximum fruit drop takes place during August-September under Tarai condition of Uttarakhand. The extent of retention and causes of fruit dropping in bael cultivars Narendra Bael 4, Narendra Bael 5, Narendra Bael 7 and Narendra Bael 9 was studied and the highest fruit drop was observed in Narendra Bael 7 (99.91%). The maximum drop (45.05%) was recorded one week after fruit setting (second week of July) in all the cultivars associated with wind speed and air temperature; whereas, relative humidity showed negative correlation with fruit drop.

Management

Spray of boric acid @ 0.1 per cent was most effective in reducing the fruit drop.

b) Fruit Cracking

Fruit cracking is found in some varieties of bael ie., NB 5(36.30%), NB 7(15.63%), NB 9(44.47%), Pant Aparna (39.47%), Pant Urvashi (34.05%), Pant Shivani (32.25%) and Pant Sujata (43.40%) and emerged as serious problem of bael cultivation under hot arid ecosystem of Rajasthan.

Management

For control, the fruits should be plucked in January and artificially ripened by treating with 1000-5000 ppm ethrel and storing them at 30°C for 18-24 days.

❐❐❐

14

Tamarind

a) Alternate Bearing

Tamarind has a tendency towards alternate bearing and this is one of the most burning problem, since it renders tamarind cultivation less remunerative to the growers. The term denotes yield variation in alternate years i.e., a year of optimum or heavy fruiting followed by a year of little or no fruiting. Some trees exhibit strong biennial bearing characteristics while some trees exhibit irregular bearing habit. The term irregular bearing implies that cropping does not follow a systematic pattern i.e., some trees bear heavy yield once in three to four years. Alternate or irregular bearing habit is a genetic factor and this trait can be identified among the tamarind trees right from the initial years of fruiting.

Management

A genotype which shows regular bearing characters would be desirable.

❑❑❑

15

Jamun

a) Flower and Fruit Drop

In Jamun, the flower and fruit drop starts just after opening of flower and continues upto maturity. About 65 per cent of flower and fruits drop in the first five weeks. Only 12-15 per cent flowers reach maturity. The flower and fruit drop are found at 3 stages. The first drop takes place during bloom or shortly thereafter. This proves to be the heaviest drop as about 52% of the flowers drop off after 4 weeks from flowering. The second drop starts about 35-40 days of full bloom and apparently there is no difference between the developing and aborting fruits. The third drop takes place after 42-50 days of full bloom and continues till 15th July.

Management

The extent of flower and fruit drop in jamun may be reduced by two sprays of 60 ppm GA_3, one at full bloom and the other at 15 days after initial setting of fruit.

❑❑❑

16

Durian

a) Water Core

Due to this disorder, the flesh areas appear water-soaked resulting in deterioration. The affected areas get decayed faster than unaffected areas. Water core is caused by rain just before harvesting. The excessive accumulation of water through the roots during maturation time of fruits is another cause behind this problem.

Management

This disorder can be corrected by avoiding excessive watering.

b) Uneven Fruit Ripening

It is a problem that has been observed in durian fruits which ripen unevenly. In such circumstances, a portion of the aril remains hard, leathery, whitish in colour, odourless and tasteless. A higher incidence of this disorder may be observed in larger fruits than smaller fruits. Incidence and severity are related to pre-harvest factors. It may be related to a combination of factors such as nutrition, water availability and environmental conditions. There is a close association of imbalances in mineral nutrition with this physiological disorder.

Management

To control this disorder, apply calcium nitrate @ 2 kg per tree one month before harvest.

c) Flesh Burn

Flesh burn is characterized by browning of the flesh. The disorder degrades the quality of fruits which fetch poor price in the market.

Management

It can be managed by the application of borax @ 10g per tree every year. The appropriate time for borax application is during flowering.

d) Chilling Injury

The black discolouration of surface, especially the groove between thorns is a peculiar symptom of chilling injury in durian. In such a condition, the fruit fails to ripe due to loss of ability to convert starch into sugars. Almost all durian cultivars are damaged by chilling injury due to storage at 5°C (40°F) for one week or 10°C (50°F) for two weeks.

Management

The incidence can be controlled by storing the fruits at safe temperature.

❑❑❑

17

Litchi

a) Poor and Irregular Fruiting

Poor fruiting/ non fruiting/ irregular fruiting are the problems in litchi growing area of the country. It has been observed that capacity of litchi shoots to bear is varietal in nature, while some bearing terminals of current year are more productive (70-95%) in the next year. Non-adoption of good management practices lead to irregular bearing even in regular bearing cultivars. Type of soil coupled with management practices also found to be associated with irregular bearing or alternate bearing. Late harvesting and severe pruning and training operations may lead to non bearing in that particular year of operation.

Usually, heavy crop is followed by a light crop in some varieties, while in some varieties production is quite consistent every year. In some cases, at the time of bearing panicle/ inflorescence, trees start giving vegetative flush and that particular year becomes off year, while in some cases, only part of the tree bear long panicle with leafy growth at the base, resulting only few flowering terminals and fruit set is also very poor.

Management

- Varieties chosen for planting must be regular in bearing, apart from other desirable characters.
- On the basis of age of the trees, the commercial orchards should be properly fed with balance nutrition immediately after fruit harvest and irrigation at right time through right method, particularly at aril development stage.
- The plant protection measures to control major pests and diseases should be followed.

- Proper pruning and training at the time of harvesting and just after harvesting is needed, While harvesting the fruit bunches, the branches having 8-10 inches length should be plucked to give rise strong and healthy flush to bear fruits in next year.
- Late pruning and training should be avoided. The tree should be trained to give canopy shape in a semi-circular manner or an open umbrella shape.

b) Flower and Fruit Drop

Sometimes even after profuse flowering and fruit set, poor yield has been observed. Because, litchi trees suffer a heavy flower and fruit drop between flowering and fruit maturity. Only a small proportion of it (2-18%), is carried to maturity in different cultivars. The flower and fruit drop is thought to be due to failure of fertilization, embryo abortion, nutrition and hormonal imbalance and external factors like high temperature, low humidity and strong westerly winds as well as due to fruit borer and heavy mite attack. The young bearing tree suffers fewer drops than the older trees. Studies revealed that maximum fruit drop during the first fortnight after fruit set and at harvest. The retention varies between 3.0-39.6% among the cultivars. The specific relationship between auxin production in developing fruit and abscission in litchi has also been taken into consideration.

In litchi, male functional flowers drop after pollen formation. The heavy fruit drop has been observed at different stages due to the formation of abscission layer. Although fruit drop continues up to maturity, but most of the flower and fruit drop occur during the first month after the pollination. The initial fruit set in litchi is very high but a very small proportion finally mature. The premature fruit drop commences soon after fruit set and continues till fruit maturity.

Management

- High fruit retention in litchi can be ensured by controlling winter shooting, flowering period and flower quality.
- Timely treatment of plants for strong and healthy shoot production delays flowering phase and increases the female flower ratio and finally the fruit setting per cent.

- Honey bees are the main pollinating agent in litchi. Increasing bee population in orchard ensures better pollination and fertilization which increases the fruit set and retention.
- It is advised not to do any type of spray at the blooming stage of the orchard.
- Spraying of ZnSO4 @ 0.2% at 30 days before panicle emergence induces healthy inflorescence leading to more fruit set.
- Foliar application of 0.2% Boric acid, 2-3 times during the period of fruit growth and development enhance fruit retention, minimize cracking, improve fruit colour, sweetness and enhance maturity leading to increased percentage of quality production.
- Treatment with growth regulators like NAA at 20-30 ppm, GA_3, at 20-25 ppm, 2,4-D at 10-20 ppm are effective in minimizing fruit drop, when sprayed on panicles, before the flower opening.

c) Fruit Cracking

In India, the development of litchi coincide with rising atmospheric temperature and low soil moisture and relative humidity that makes more congenial for skin to crack. Temperature higher than 38°C in combination with relative humidity lower than 60% was very favourable for cracking of litchi and causes 10-25% crop loss. This may occur due to varietal characters, orchard soil management, inappropriate levels of water at maturity stage, light, mechanical injuries, temperature and micronutrient deficiency. Cracking in litchi is most common in China and India, the two largest litchi producers in the world. Fruit growth in litchi consists of two distinct phases: growth of the pericarp (skin) and seed, followed by growth of the aril. Cracking injury may vary from short cracks, generally skin deep to larger ruptures extending to most of the length of the fruit. In litchi, fruit cracking may be initiated as microcracks during it's growth and development on exposure to low relative humidity and hot dry winds. Microcracks formation is further triggered by the rapid expansion of the aril at the initial stage of fruit development which intensified with maturation.

Factors affecting fruit cracking

There are many factors associated with fruit cracking problem which includes morphological, genetical, environmental and physiological aspects.

(i) Fruit Composition

Litchi fruit consists of a glossy and brown single seed surrounded by a sweet-acid, crispy, white to cream-coloured, juicy and translucent aril. Variation in relative proportions of reducing and non-reducing sugars of pulp could be important in fruit cracking. Sucrose, fructose and glucose are found to be the major sugars in litchi. Total sugars in physiologically mature fruit may range from 55.9 to 61.4% which is mainly represented by reducing sugar (> 70%). The relative ratios between these sugars may be different in various cultivars, stage of maturity and enzyme activity. Cultivar Gola with high per cent of fruit cracking also had highest fruit weight, reducing sugars, specific gravity and TSS but the least non-reducing sugars.

(ii) Skin morphology

Comparatively cultivars with thick skin and fewer tubercles per unit area of skin were less susceptible to cracking. The skin strength in litchi fruit is an important and desirable characteristic because an imbalance in cell turgidity and extensibility could lead to fruit cracking.

(iii) Turgor Pressure

Cracking of fruit is basically caused by internal stresses due to turgor pressure and differential growth between the parenchymatous tissue and the skins. It was hypothesized that a balance between turgor pressure, structure and elasticity of skin could prevent fruit cracking in litchi and that fruit cracking occurs only when aril produces more turgor pressure against the skin or reduces the structure and elasticity of the skin.

(iv) Fruit growth characteristics

Fruit growth exhibited two stages in terms of the differential growth rate of pericarp and aril. Stage I (slow growth phase) continues up to 52 days after anthesis (DAA) and was mainly characterized by pericarp and seed growth, while Stage II (rapid growth phase) was markedly characterized by aril growth. Fruit cracking occurred in a period of 66-80 days after anthesis.

(v) Changes in mineral composition

The decline in pericarp Ca and B concentration prior to aril growth can be accounted for the cracking incidence. Boron and calcium contents in pericarp of uncracked fruits were significantly higher than those of cracked fruits.

(vi) Enzymatic activities

The changes in pectin methyl esterase (PME) and polygalacturonase (PG) activities during the early stages of fruit development act as key factors in fruit softening which in turn, could change the properties of pericarp cell wall through regulation of pectin metabolism. The activities of the wall-metabolic hydrolases, viz. PG, cellulase and PME and wall-bound peroxidase (POD) and polyphenol oxidase (PPO) were higher in the pericarp of susceptible cultivar. Cracking resistant varieties had a higher levels of structural calcium and galacturonans with lower activity of PME and ionically wall-bound POD.

(vii) Weather

Climatic factors, such as drought, high temperatures and excessive rains play a dominant role in inducing fruit cracking. Temperature and exposure of fruits to sunlight as well as atmospheric wind and low humidity surrounding the fruit have been considered as possible causes of cracking. Continuous rain or heavy irrigation after a dry spell favours high absorption of water by fruit and aggravates the incidence of fruit cracking. Inadequate moisture during the early fruit growth coupled with high temperature affect cell division, makes the skin hard and inelastic and it may get cracked subsequent to increased internal pressure exerted by the growing aril following irrigation.

(viii) Nutritional status of soil and plant

The contents of leaf Ca, Mg and B, pericarp N/K ratio, soil N/K ratio, organic matter content and exchangeable Ca of soil influence cracking. Soil nitrogen content was positively correlated with fruit cracking, but was the reverse in case of leaf nitrogen concentration. Cracked fruits contain higher N, P and K but lower calcium and zinc than normal fruits. Low soil phosphorus and magnesium correlates with fruit cracking. Calcium played a dual role in relation to litchi pericarp cracking, i.e. structurally and non-structurally via spongy tissue formation, a process essential for pericarp extension during aril growth.

(ix) Hormonal balance

Basipetal movement of auxin regulates calcium uptake by fruit at their peak metabolism period and this might be accounted for the difference in calcium content between cultivars. Abscisic acid (ABA) concentrations were higher in the pericarp, seed and aril of cracked fruit than in normal fruit, while the concentrations of gibberellin (GA_3) were higher in seeds at the critical period of cracking.

(x) Genetics / Cultivars

Resistance to fruit cracking is independent of calcium contents in the aril and pericarp, thus indicating the presence of a genetical link. It was reported that early ripening cultivars were more susceptible than later ones. In general, cultivars, which have relatively thin skin, few tubercles per unit area and rounded to flat in shape are less prone to crack.

Management

(i) Calcium

Spraying thrice with $Ca(NO_3)_2$ 100 mmol/litre at full bloom, 19 and 39 days after full bloom reduced fruit cracking.

(ii) Boron

Spraying boric acid at 1.0% during the developing stage of the fruits were highly effective in minimizing fruit cracking.

(iii) Zinc

Spray of $ZnSO_4$ (1.0%) during first week of May at pit hardening stage minimized fruit the cracking.

(iv) Plant growth substances

Growth regulators, such as NAA (20 mg/litre), gibberellic acid (40 mg/litre), 2,4-D (10 mg/litre), 2,4,5-T (10 mg/litre) and Ethephon (10 mg/litre), were found to be beneficial in lowering the incidence of fruit cracking.

(v) Irrigation

Drought or deficit irrigation promotes fruit cracking in litchi. Light but frequent irrigation that maintain a congenial microclimate within and around the plant has been found to minimize this problem. Though water requirement in litchi vary with plant age, size as well as seasons and locations, but on an average litchi plant requires 600-800 mm water. Irrigation gave best result in checking fruit cracking, when bearing trees are irrigated at 100% evapotranspiration replenishment. Irrigation at 20% pan coefficient in conjunction with sprinkler irrigation or placement of dessert cooler was highly efficient in control fruit cracking.

(vi) Mulching and bagging

Mulching, either plastic or organic mulch or along with three irrigations was effective in checking fruit cracking. Bagging reduce the incidence of fruit cracking by altering the microenvironment for fruit development. Litchi fruit bagged with cellophane or fabric bags from 15 days after full-bloom until harvest showed better skin colouration.

(vii) Integrated management of fruit cracking

Micronutrients (0.8% zinc sulphate and 0.4% boric acid) combined with 10 ppm GA_3 was effective in reducing fruit cracking. Calcium nitrate (0.5-1%) along with boric acid (0.4-0.8%) or when combined with irrigation (100 litres) also showed a pronounced effect in controlling fruit cracking. The integrated approach which consists of mulching the tree basins with fallen litchi leaves, spraying of different nutrient elements (B and Zn) and Kaolin, irrigating the orchard at 40% depletion of available soil moisture, sprinkler irrigation, bagging of fruit bunches and protecting the fruits with shade nets at 30 days before harvesting was effective in reducing fruit cracking.

d) Sunburn

Sunburn also known as lesion browning or pericarp necrosis is a serious problem in litchi. Climatic factors and cultivars in particular growing areas are determinants for incidence and severity of sunburn disorder. The damage caused due to sun burning which occurs upto 0.9-19.13% in different varieties. Apart from environmental factors, varietal, hormonal, nutritional and soil moisture factors are associated with this disorder. This disorder is physiologically related with PPO (Poly-phenol oxidase) activities and it also varies with cultivars. Sunburn is pronounced in ill managed orchards having sandy or sandy loam soils or light soils receiving/exposed to high temperature (>40°C) and very less RH (<50%). It is a type of direct thermal injury and in case of higher temperature, the tissue coming in contact/exposure gets sunburnt/ sunscalded. Sunburn problem is also seen more in early ripening cultivars. Fruits on shaded branches suffer less damage than those more exposed to sun. Lower translocation of calcium in the pericarp region is also found to favour sunburn disorder.

In case of sunburn, light brown blotches appear on the portion of the fruit skin facing direct sun rays. In severe cases, more than half of the surface area becomes discoloured and blotchy light brown. The blotches become intense in few days and the blotchy area dries up blocking of the aril growth and finally destroying the market value of the fruit. The symptoms appear more on the south west side than the north-east side as on the latter side fruits remain almost in shade except during early hours of the day, which is not harmful. Under similar climatic conditions, the bearing of the fruit may suffer with this disorder due to the position of the fruit on the tree.

Management

- Irrigation at regular interval during the fruit growth and ripening stage reduces sunburn.
- Planting wind breaks around the orchard provides protection from desiccating hot winds, thereby resulting in low sunburn.
- Irrigation through sprinkler system during hot hours increases humidity, cools the orchard atmosphere thus, decreases the incidence of sunburn.
- In light and sandy soil, only light irrigation with increased frequency (4-5 days interval) is found beneficial.
- Feeding trees with sufficient quantities of organic manures particularly compost, FYM, cakes, green manure, vermi-compost along with applied irrigation at regular interval during fruit development and ripening stage have been found useful.
- Planting of maize or sugarcane around the litchi orchard and daily irrigation in such border crop creates congenial microclimate, which reduces sunburn.

e) Underdeveloped Fruits

This disorder is causing considerable loss due to poor quality fruits or ill developed fruits. The fruit size remains smaller than the normal and juice content or aril development is also very poor. This type of fruits remain in glossy pale green colour for a longer period and then only dull reddish colour appears and just after, the problem of prematurely fruit drop also starts. In fruit growth of litchi, seeds develop initially at a high rate, followed by membranous mesocarp and aril, which grow very fast towards the later stage of fruit development. During these phases, sometimes, it happens that there is no aril development or aril development incomplete with portion of testa uncovered also give rise to the development of retarded/ underdeveloped fruits. Lack of proper fertilization because of less visits of pollinators may lead to improper fruit set and retarded fruit growth. Insect-pests attack and physical jerk also cause retarded or underdeveloped fruits. Unavailability of moisture during development stage of fruit is also one of the causes for retarded/ underdeveloped fruit. It is more prevalent in poorly managed orchards having poor pollination.

Management

- Proper nutritional management of orchard.
- Establishing enough number of honey bee colonies during flowering to fruit set stage increases fruit set with better fertilization leading to normal and healthy fruit growth.
- Spraying plain water in early morning hours of the day during the advanced stage of growth and development have been found highly effective in better growth of fruit and quality.
- Arrangements of sprinklers system of irrigation has also been found to be effective in reducing this disorder.

f) Black Spot

This disorder has been observed in existing plantations mainly in the vicinity of urban areas. It is caused due to deleterious effect of smoke fumes which contain sulphur dioxide, acetylene and carbon dioxide as found in case of mango. The symptoms of black spot become apparent, when the fruit attains certain size and remain in developmental phase i.e. from March end to mid May. The first symptom appear as the development of a small etiolated area at the distal end of the fruit against the normal green colour of the fruit pericarp, which gradually spreads, turns nearly black and covers the distal end completely. The infested fruits show discolouration and overall slow pace of development resulting in very poor development of aril, seed, size and colour.

Management

- The disorder can be reduced by applying proper nutrition and irrigation particularly during fruit growth and aril development stage.
- Spraying with micronutrients like zinc ($ZnSO_4$ @ 0.2%) a month prior to flower panicle initiation and boron (boric acid @ 0.5%) during fruit growth stage can reduce the problem.

g) Pericarp Browning

Pericarp browning is the first visual symptom of fruit decline. Browning that occurs during the first few days after harvest is usually caused by dehydration of the pericarp. This problem has been reported in several litchi producing areas and involves the appearance of necrotic spots which may expand and affect a significant portion of the fruit. Fruits start to brown once they lose a few per cent of the harvested pericarp fresh weight. Below 50 per cent of its initial fresh weight, the pericarp is entirely brown. Browning of the pericarp occurs at ambient temperatures of 20-30°C within 24 hours of harvest. Water loss (desiccation) of litchi results in brown spots on the bright-red shell (pericarp). Other factors also cause the fruit turn to brown, includes sliding the fruit down a rough picking bag, dropping fruit from short heights, microbial and insect attack and extremes of temperatures. In short, anything likely to accelerate cell breakdown is likely to increase fruit browning. This occurs as a result of damage (due to causes unknown) to the parenchymatous cells of the thin wall of the mesocarp. This damage speeds up a characteristic hypersensitive reaction, leading to necrosis of these cells, which then spreads to the epicarp and endocarp. Skin-browning is associated with an increase in polyphenol oxidase and peroxidase enzymes activity and with ascorbic acid oxidation. Losses may be considerable, although the aril is not affected; the fruit is not suitable for marketing. In extreme cases, the pericarp may crack. Through the use of fungicides and refrigeration, litchi fruits can be stored up to 30 days. However, when they are removed from storage, their shelf life at ambient temperature is very short due to pericarp browning and fruit rotting.

Management

- Packing fruit into moisture-proof (plastic) bags can substantially reduce water loss and slow the rate of browning.
- Cool temperature storage also slows browning. Low temperatures slow evaporation as well as respiration and probably slow tissue senescence. It has been reported that fruits treated with polyamines, suspected anti-senescence agents, then wrapped and stored at 5°C, had lower membrane permeability and less browning than controls.
- A controlled atmosphere of 3 to 5 per cent O_2 and 3 to 5 per cent CO_2 has also been shown to slow water loss. Fruits stored under such an atmosphere for 30 days at 1°C lost only a quarter of the water lost by the control. However, the mechanism of the response is not clear. Such an environment may affect the metabolism of the fruit as well as that of the pathogens.
- Sulphur dioxide fumigation effectively reduces pericarp browning, but some countries do not permit due to concerns over sulphur residues in fumigated fruit. Thus, sulphur-free post-harvest treatments that maintain fruit skin colour are increasingly important.

❑❑❑

18

Mangosteen

a) Gamboge Disorder

Affected fruits exude yellow latex which is responsible for the discolouration and change in taste. In fact, the yellow resin is naturally produced by several organs which consist of branched canal-like secretory ducts. They are found in the exocarp, mesocarp and endocarp, aril of the fruit, flower, stem and leaf. The biggest secretory ducts are located in the endocarp which continues from the fruit stalk to the fruit. The bitter yellow resin permeates into the aril as a result of broken latex secretory ducts caused by excessive rain, winds, careless fruit handling which induce physical damage to the fruit pericarp. Latex secretory ducts are usually broken at 5-15 weeks after anthesis during the rapid fruit growth period. The seed and aril growth rates are reported to be faster than the pericarp causing a building up of internal pressure to the endocarp. The resultant pressure causes the secretory ducts to break especially when the epithelial cells of the ducts are weak. However, this problem is rectifiable by application of adequate calcium nutrient to mangosteen trees. Calcium as one of the important components of the middle lamella in the form of calcium pectate that binds together the continuous cell walls is a major factor in withstanding such pressure.

Management

- Application of dolomite fertilizer at rate of 18-24 tonnes/ha will improve calcium content in the exocarp and leaves and it effectively reduces gamboge disorder (GD).
- Applications of $CaCl_2$ spray effectively reduced yellow spots either on the pericarp or the aril of mangosteen.
- Meanwhile, a mild soil water stress of approximately -70 kPa should be maintained during pre-harvest.

b) Transparent Flesh Disorder

Transparent Flesh Disorder (TFD) of mangosteen fruit shows as a water soaking of the flesh. The incidence of TFD is caused by excess water during the pre-harvest stage of fruit development, where water uptake was driven by the difference in osmotic potential between that of rain water and that of the affected fruits. When water content in fruit is high, this leads to a breakdown of apoplast or symplast compartmentation which is exhibited as the occurrence of TFD. It was found that the threshold of TFD incidence of mangosteen fruits subjected to excess water is around 9 weeks after bloom.

Management

Transparent flesh disorder incidence could be avoided by maintaining mild soil water deficit around -70 kPa during pre-harvest.

❑❑❑

19

Loquat

a) Creasing

Creasing is caused by the formation of abscission zone between the fruit base and the peduncle. The incidence of skin creasing coincides with spring flushing and with the maturation of spring shoots, the incidence of creasing decreases. Although, spring flushing might be the main cause of creasing before harvest, it may also be related to a complex of pre-harvest factors, including potassium nutrition.

Management

Pre harvest sprays of gibberellic acid have been found to reduce creasing.

b) Internal Browning

Internal flesh browning followed by tissue breakdown is enhanced by higher temperatures and longer durations of storage. Elevated CO_2 concentrations (>10%) may induce internal flesh browning and skin brown spotting.

Management

Storing the fruits at optimum temperature and CO_2 concentrations can prevent internal browning in loquat.

c) Russeting

It is the appearance of corky, roughened, brownish or greyish areas on loquat fruits. It is characterized by number of small cracks which are often concentric and pre-harvest skin blemishes (russeting) may appear on developing fruit. It's severity depends on cultivar, season and microclimatic conditions. There are several possible causes for fruit russeting, including cool wet weather, frost damage, spray damage, etc. The first symptoms of russet appear 30-42 days after fruit set, during the final stages of cell division. It develops during 54-76 days of fruit development i.e. at the beginning of the cell enlargement phase.

Management

- Proper chemical concentrations should be used for spraying loquat. Spray only under proper conditions by using the correct equipment.
- Proper water management practices should be followed to control growth and irrigate at fixed intervals and reduce the frequency of watering, when necessary.
- Do not remove leaves excessively as leaves create a micro-climate for the fruit which helps prevent russeting.

d) Sunburn

Sunburn, or "Purple spot", is responsible for much fruit loss in hot regions with long summers. During winter if the heat is high, it may also result in sun burned fruits. To produce high quality fruits, the fruit clusters may be bagged so as to protect them from being damaged by sunburn (purple staining of the skin).

Management

Bagging of the fruit can protect them from this blemish.

❑❑❑

20

Noni

a) Crooked Fruits

In noni, crooked fruits are developed due to damage of fruit eye by insect, disease or mites. It hampers proper development of the fruit and they get curved in damaged eye side and deform the fruit shape. The intensity of such fruit increases during rainy season. The incidence of crooked fruits is very less and losses due to this in noni cultivation are very negligible.

Management

Proper sanitation and control of pests also reduce the incidence of crooked fruits.

b) Fruit Deformity

Noni is a medicinal plant and industry requires fruits of uniform size and shape for easy processing. But, due to climatic or nutritional factors, some of the fruits produce erratic shaped fruits with out-growth. It also hampers grading, packaging and storage practices.

Management

The exact factors for this disorder are still unknown but proper nutrition and regulated irrigation reduces incidence of such fruits in noni crop. The CARI, Port Blair has identified a genotype GAH-1 which produced 80-90 per cent uniform fruit with minimum incidence of fruit deformity.

❑❑❑

21

Apple

a) Water Core

Water core is promoted by larger fruit, high leaf to fruit ratio, high fruit nitrogen and boron, low fruit calcium, excessive thinning, high light exposure, ethrel sprays and girdling of trunks and limbs. Another type of water core which is unrelated to maturity may occur during unusually hot weather. This water core is found more on the exposed side of the apple and may be associated with sunscald. If moderate to severe, results in reduced storage life of the apple, as internal breakdown develops in affected areas. A pre-harvest disorder resulting in water soaked regions in the flesh, hard and glassy in appearance, visible externally when very severe. Water soaked areas are found near the core and around the primary vascular bundles but may occur in any part of the apple or involve the entire apple. If symptoms are mild to moderate, they may disappear completely in storage. However, when watercore is severe, internal breakdown can develop. The water soaked appearance of watercore affected fruit results from the accumulation of sorbitol rich solutions in the intercellular spaces.

Sorbitol is the carbohydrate source translocated into the fruit from the tree. Sorbitol must be converted to fructose by the apple fruit. The reason for accumulation in the intercellular spaces is not known. It is speculated that sorbitol may be translocated to the fruit faster than it can be assimilated perhaps due to an inability of the apple tissue to convert sorbitol to fructose. Comparative studies have shown differences in vascular tissues between susceptible and resistant varieties. The browning and breakdown that results from severe watercore is likely due to reduced gas diffusion in the affected tissue and may accumulate of ethanol and acetaldehyde.

Management

Experiments have shown that water core is reduced by raising the calcium content of the fruit and is increased by high applications of nitrogen and boron. The best way to minimize water core in apples is to avoid late harvesting.

b) Storage Scald

Storage scald has been a serious concern for apple growers as long as apples have been stored and marketed commercially. Susceptibility to this storage disorder varies with the variety of apple, environment and cultural practices. Incidence and severity of storage scald is favoured by hot, dry weather before harvest, immature fruit at harvest, high nitrogen and low calcium concentrations in the fruit and inadequate ventilation in storage rooms or in packages.

Irregular brown patches of dead skin which can become rough when severe, develop within 3 to 7 days upon warming of the fruit following cold storage. The warm temperatures do not cause the scald but allow symptoms to develop from previous injury which occurred during cold storage. Symptoms may be visible in cold storage, when injury is severe. In this case, the symptoms intensify upon warming the fruit. Scald is usually not evident until after 3 months of storage. Storage scald can involve only the skin or extend 6 mm into the flesh depending on the variety. Scald can be more severe on the greener side of the fruit. There continues to be much debate about the cause of storage scald in apples, but most agree that storage scald is a type of chilling injury. The general theory is that alpha-farnesene, a naturally occurring volatile terpene in the apple fruit, is oxidized to a variety of products. These oxidation products result in injury to the cell membranes which eventually result in cell death in the outermost cell layers of the fruit. Ethylene promotes the formation of alpha-farnesene and oxygen is required to oxidize alpha-farnesene to conjugated trienes.

Management

Mineral oil wraps are shown to absorb the volatile alpha-farnesene and ventilation also prevented accumulation of alpha-farnesene in the storage room air. Today, the most common method used to control scald is application of an antioxidant immediately after harvest. Diphenylamine (2000 ppm) is commonly used. Application of $CaCl_2$ (2-3%) two weeks before harvest is very effective for controlling scald effectively. Antioxidants should be applied within one week of harvest for maximum control. Low oxygen controlled atmosphere storage can provide a non-chemical control method in some cases. Oxygen concentrations between 0.5 and 1.0% can significantly delay the development of scald, perhaps upto 10 months.

c) Sunburn (Sun Scald)

It can occur to some extent in all apple growing regions but is especially a problem where temperatures are high and skies are clear. Found more frequently on fruit on the southwest quadrant of the tree. Heavy crops that cause branches to bend over mid-season can increase sunburn incidence as a sudden exposure to heat and sun promotes sunburn development. Water stress can increase the incidence of sunburn.

Initial symptoms are white, tan or yellow patches found on the sun exposed side of the fruit. With severe skin damage, injured areas can turn dark brown on the tree. White, tan or yellow patches often turn brown within few weeks in cold storage. During unusually hot weather following cool or moderate conditions and especially when accompanied by water stress, injury to the skin and flesh can occur. Injured cortex tissue is brown and firm and may become spongy and sunken. Fruits exposed to the sun after removal from the tree, either on the orchard floor or in field bins, can develop severe sunburn. Sunburn results from heat stress to the fruit leading to injury of the affected cells. The temperature of the surface of the fruits can be as much as 18°C (32°F) above air temperatures when the fruit is exposed to solar radiation and 8 to 9°C (14 to 16°F) warmer than the shaded side of the fruit. The transpiration of water from the apple fruit helps to cool the fruit while it is attached to the tree. Upon harvest from the tree or during periods of water stress on the tree, fruit surfaces and flesh can increase to much higher temperatures resulting in greater sunburn. The greatest problem occurs, when fruits are suddenly exposed to high temperatures and intense sunlight, such as when the weather rapidly shifts from being cloudy and cool for many days to sunny and warm. A gradual increase in temperature and solar radiation could allow the fruit to adapt; whereas a sudden change does not allow for adaptation and often results in injury.

Management

The best method of control is to avoid sudden exposure of fruit to intense heat and solar radiation. Proper tree thinning and pruning are critical. Summer pruning must be carefully accomplished to avoid excessive sunburn. Overhead sprinkling and white washing of trees have both been tested with mixed success in reducing sunburn. Avoid water stress in cropped orchard trees to reduce heat stress. Once injury has occurred, there is no treatment to prevent browning in storage. Careful sorting to remove affected fruits upon packing is the only solution once the injury has occurred. Injury will be easier to spot after a few weeks of storage because of the change in colour from yellow to brown.

d) Bitter Pit

Bitter pit reduces the fresh market quality of fruit. It was first described in Germany and recognized as an abiotic disorder found in all areas of the world where apples are grown. This physiological disorder is influenced by climate and orchard cultural practices. Small brown lesions of 2-10 mm in diameter (depending on the cultivar) develop in the flesh of the fruit. The tissue below the skin becomes dark and corky. At harvest or after a period of cold storage, the skin develops depressed spots on the surface. These most often start to appear as water soaked spots on the skin near the calyx. These spots generally turn darker and become more sunken than the surrounding skin and are fully developed after one to two months in storage. Fruits located on vigorous, leafy, upright growing branches have a greater potential to develop bitter pit than does fruits that develop from spurs or on horizontal wood near the tree's main frame.

Young trees that are just coming into bearing are the most susceptible. Immature fruits are more susceptible to bitter pit than fruits harvested at the proper harvest maturity. Older trees, which are less vigorous and produce larger crop loads, reduce their susceptibility to bitter pit, except in the very sensitive cultivars. Initiation of symptoms may begin four to six weeks after petal fall, when affected tissues have a higher rate of respiration and ethylene production. A mineral imbalance in the apple flesh develops with low levels of calcium and relatively high concentrations of potassium and magnesium. Low levels of calcium impair the selective permeability of cell membranes leading to cell injury and necrosis. Other explanations for the cause of bitter pit include dissolution of the middle lamellae by oxalic and succinic acid and changes in proton secretion and potassium permeability.

Management

Cultural practices that reduce the incidence of bitter pit are annual bearing, moderate tree vigour, smaller fruit size, calcium sprays, summer pruning and harvesting mature fruits. Avoid nitrogen and magnesium summer sprays, fertilizer treatments that result in lowering the soil pH or induce excess vigour and fluctuating soil moisture. Early thinning and over thinning can increase bitter pit. Do not over prune, which would result in a light crop and large fruit. Summer sprays of 0.2 per cent calcium chloride ($CaCl_2$), calcium nitrate [$Ca(NO_3)_2$] and/ or a postharvest dip in a calcium solution are recommended. Both $CaCl_2$ and $Ca(NO_3)_2$ summer sprays are effective in reducing the incidence of bitter pit.

e) Internal Browning

This disorder is associated with later harvested, large and over mature fruits and with CO_2 concentrations in storage. Higher concentrations of CO_2 result in greater incidence and severity of internal browning. Incidence and severity of this disorder varies season to season and orchard to orchard. The reasons for this variability are unknown but some have suggested an association with cool, wet weather and high nitrogen fertilization. Because, internal browning is not detectable externally, except in very severe cases, affected fruit can be discovered by buyers or consumers, thereby affecting future confidence in the product. Although fruit susceptibility varies season to season and orchard to orchard, considerable losses of controlled atmosphere-stored fruit can result from this disorder.

Brown discolouration in the flesh, firm but moist, usually originates in or near the core. Brown areas have well-defined margins and may include dry cavities resulting from desiccation. Symptoms can range from a small spot of brown flesh to nearly the entire flesh being affected in severe cases. When the entire apple is affected, a margin of healthy, white flesh usually remains just below the skin. Symptoms develop early in storage and may increase in severity with extended storage time. Evidences indicate that internal browning occurs as a result of CO_2 injury to the apple. Injury incidence and severity increases with increasing concentrations of CO_2 in the storage atmosphere. Variability in susceptibility of apple varieties and in apples of different maturities may be due to anatomical differences (cell size, size of intercellular spaces) rather than biochemical differences. Later, harvested fruits generally have greater susceptibility to internal browning and it has been shown in many fruits that diffusivity of CO_2 decreases as fruits mature and ripen. A reduction in the ability of CO_2 to diffuse out of the fruit would result in higher internal CO_2 concentrations and thus, more chances of injury. Internal browning can be more severe in fruits, when water core is also present.

Management

Avoid harvesting over mature fruits. Harvest at the optimum maturity, especially for controlled atmosphere storage. Assure good air circulation in storage rooms to prevent pockets with higher CO_2 concentrations. CO_2 can develop to damaging levels in air storage or marine containers if fruit temperature is 3°C or warmer and ventilation is not good. Proper temperature management and good ventilation will prevent build up of CO_2. Avoid heavy wax coatings and thoroughly and rapidly cool the fruits after waxing and packaging.

f) Cork Spot

The initial symptoms of this physiological disorder appear as small blushed area on the skin of the fruit above the affected brown spot. The affected tissue is usually much harder than the healthy tissue. Boron and calcium deficiencies are occasionally found responsible for development of cork spot.

Management

Proper nutrient management, especially boron and calcium helps in preventing this disorder.

g) Russeting

Apple fruit russeting is undesirable because, it lowers the market value of the fruit. Cultivar Golden Delicious provides the world wide example of this depreciation. Russeting is a surface blemish in which normally smooth cuticle and epidermis have been replaced by a rough corky tissue or periderm. Russeting is caused by damage to epidermal cells during their intensive division shortly after petal fall. As cells develop in this area they push upward and become exposed to the surface as the fruit matures. Russeting of apples in a humid environment is a major concern of the fruit growers. The water imbalance is considered critical in the formation of russet as it develops under conditions of high humidity, occasional rain or dew. High turgor pressure in epidermal cells due to diffusion of rain or dew through the cuticular cracks on fruits exposed in the sun may cause bursting of skin and thus, russeting is more commonly noticed on exposed fruits than on fruits in shade. Frost during blossom or at the early fruit let stage may also caused russeting.

Several fungi including powdery mildew causing fungi are also involved in its development. The apple cultivars, which have thin cuticle, are more susceptible to russeting. Frost during the blossom or at the early fruit formation stage may also cause russeting. Russeting leads to rupture of the fruit skin and development of cracks.

Management

Spraying 0.7% wettable sulphur plus 0.1% borax plus 0.2% kaolin @ 1500 litre/ha at petal fall and every 8-10 days thereafter until the fruit hangs downwards. Selection of less susceptible clones and adequate irrigation, manuring and effective pest management can also reduce russeting.

h) Fruit Drop

In apple, there are four waves of fruit drop at about 12-14 days interval. The first fruit drop occurs shortly after petal fall. It is by far the largest and tends to overlap with second drop. The first two drop occur in all varieties under all conditions and have been found more pronounced in some varieties, such as Delicious. These two drops are collectively called "June drop". The early drop is considered natural and is due to lack of pollination and fruit competition. Moisture stress and environmental conditions cause the June drop. These two drops neither cause substantial economical losses nor are controlled effectively by aritificial means. The pre-harvest drop causes serious economic loss as the full grown marketable fruits abscise before the harvest due to the reduction in the levels of auxins. The third and fourth drops are more conspicuous because the size of the dropping fruit is much larger. The process of separation has more than one phase, the first being methylation in which the action of enzyme, pectin methylesterase (PME) decreases binding by decreasing cross linkages. Pectin methylesterase is found in higher concentration near the abscission zone than elsewhere in the stem. Finally, another enzyme polygalactouronase breaks up the long chain molecules resulting in separation. At this point, the fruit is held on the tree only by vascular strands of the conducting system which are easily broken.

In fruit crops which require pollinizers for proper pollination, a desirable pollinizer(s) should be planted in an orchard in a suitable ratio. For example, Tydeman, Early Worcester, Red Gold and Golden Delicious are good pollinizers for apple. It is not only the proportion but placement of pollinizers in an orchard that holds importance. For planting 11% pollinizers in an apple orchard, every third plant in third row should be a pollinizer.

Management

Different growth regulators can be used to check the pre-harvest fruit drop. However, care should be taken because a concentration recommended for a particular variety or species may be completely useless for the same species under different agro climatic conditions. NAA @ 20 ppm, 3-4 weeks before harvest prevents fruit drop in apple. Also, the pre harvest application of ethrel @ 50 ppm has been found effective for preventing fruit drop.

i) Senescent Breakdown

With slight variations among varieties, symptoms are as follows; flesh becomes very soft, first on one side then on entire fruit. Apple is easily punctured with thumb and is easily bruised; flesh becomes dry, mealy and lacks aroma and flavour; browning of the fruit flesh and vascular strands in the flesh; skin becomes dull, dark, or water-soaked in appearance; fruit splits and open. Senescent breakdown has been aggravated by low calcium levels in the fruit, very hot and dry weather before harvest, delayed harvest, slow cooling in storage, delayed establishment of low O_2 in controlled atmosphere (CA), high storage temperature and prolonged storage.

Management

Senescent breakdown can be controlled or alleviated by drenching the apples with 2 to 4% food grade calcium chloride, rapid cooling, rapid establishment of the proper CA atmosphere and sale of the apples soon after CA rooms have been established.

j) Jonathan Spot

Jonathan spot is a disorder affecting the skin of the apple. The disorder appears as a circular spot, 2-4 mm in diameter, brownish or almost black, when associated with red skin and usually more prevalent on the highly coloured part of the apple. On green or yellow skin, the spots are lighter. Occassionally, the symptoms occur before harvest on well-matured fruit, but they usually appear after several months of storage, during which time the spots increase in size and sometimes coalesce to produce an irregular outline. The spots often, but not always, occur around the lenticels. Spotting that is confined to the lenticels is considered by many physiologists to be another form of Jonathan spot. Lesions sometimes become partly sunken, particularly if fungal infections are present. Lesions rarely affect hypodermal tissue.

Management

Prompt refrigeration after harvest, avoiding delays at ordinary temperatures. Store in CA with adequate CO_2.

k) Freezing Injury

Slight freezing injury may not be noticeable externally, or even internally, although such apples will be softer and not keep as well after defrosting. But, if severe, the surface is discolored and darkened in irregular-shaped areas and is water-soaked. In frozen condition, apple skins show network of wrinkles and, of course, apple has "hard feel." Best indication of freezing injury is found, when defrosted apples are cut. Core "dots" will be brown as well as strands in flesh if freezing has been severe. In extreme cases, flesh has solid discoloration, shades of brown or even black with water-soaked appearance. Frozen apples usually decay if left in storage. Slightly frozen apples will defrost with no apparent after-effect except slight softening, if thawed slowly. It is best not to move or handle apples when frozen because bruising injures frozen apples severely.

Management

Prevent cold "pockets" in storage by air movement and do not allow temperature to drop below 30°F (-1°C).

l) Lenticel Spot

Resembling to Jonathan spot, it is always associated with lenticels and is often caused by chemicals that come in contact with fruits. These chemicals include ammonia, sulphur dioxide, urea formaldehyde (found in fruit trays), volatile substances from ripe fruits, ethylene, or salts used in dump tanks. Ultraviolet light can also induce the disorder. The disorder does not appear to be closely related to the chemical composition of the fruit, but an increase in its calcium level by pre-harvest sprays or post harvest calcium dips is effective in reducing it.

m) Carbon Dioxide Injury

If the carbon dioxide (CO_2) level is too high during controlled atmosphere (CA) storage, both external and internal injury symptoms may occur. External injury resembles snowflake like patches which may join to form one very large patch. Internal symptoms appear as discoloured areas within the vascular bundles. The tissue may be brown and develop pockets. A noticeable aroma of fermentation may be present, when CA storage is opened, or the fruit is cut.

n) Chilling Injury

Fruits stored below their tolerance point may show a variety of symptoms depending on the cultivar, temperature and duration of storage at the adverse temperature. Symptoms may vary from browning of the skin to deep flesh browning and translucency. Dry internal cavities and flesh browning may also develop. Fruits also may have a bitter taste and smell of fermentation upon cutting.

o) Humidity Disorders

Excess or too low humidity can have a detrimental effect on fruits during storage. If the humidity is too high, the cells swell and expand to the point, where the tissue cracks. Too low humidity causes dehydration of the tissues leading to shrivelling.

p) Poor Colour Development

Nearly all red colours in fruits are due to soluble anthocyanin pigments; these arise from a product known as chromogen. The followings are the factors affecting colour development viz.,

- Ultraviolet and blue lights affect most red colouration of fruit but these photo-spectra are readily absorbed by dust and moisture in the air. The ultraviolet rays are found in higher concentrations at higher elevation and thus, at higher elevation, fruits are found to develop better red colour.

- Cool day and night temperatures or high night and cool day temperatures were favourable for red colour development but high day temperature is found to inhibit colour development even when nights are cool. The average absolute humidity favours a great amount of red colouration.
- Presence of adequate number of leaves increase sugar content of fruits and helps in the development of colour.
- Improvement in fruit colour of a dwarfing rootstock is due to the greater exposure of fruits to sun.
- In general, nitrogen inhibits proper colour development, especially when applied 2-3 months before harvest. High potassium is also found to improve red colour development.
- Fruits with better appearance often occur on the top and other exposed parts of the tree canopy. Ripened bottom apples show poor colour but good eating quality and top fruits have a better colour and poor aroma and taste.
- Trees which received a continuous moderate water supply generally produced good-coloured fruits that those grown under stress.

Management

Different growth substances recommended for improving colour of the fruits are 2, 4, 5-T, NAA and ethrel.

22

Pear

a) Bitter Pit

Bitter pit in pears is an abnormality in the cortical tissue. It develops during the later part of the growing season and is quite similar to bitter pit in apples. Patches of corky brown tissue, usually near the calyx end, occur deep in the tissue or immediately under the skin and appear externally as dark green sunken areas. Affected fruit becomes partly yellow and softens prematurely, even in storage. Tissues prone to cork is low in calcium or has a high K: Ca ratio, respires more intensely, produces more ethylene and has a higher chlorogenic acid content in the skin. High rates of nitrogen application increase the incidence of the disorder.

Management

Calcium sprays have a beneficial effect on the prevention of cork spot; late sprays are more effective than early sprays and multiple sprays are more effective than single sprays. Efficacy depends on concentration, but risk of phytotoxicity also increases.

b) Black end

This disorder, restricted to pears, is often characterized by black discolouration of the calyx end. The calyx end may become peaked or flattened; the epidermal tissue is tight and glossy. The disorder is initiated 45-60 days after full bloom. Black spots may extend to the stem end in severe cases and cracks in the core area in the calyx may be present. Affected tissue remains hard, very gritty and seldom softens upon ripening. Fruits with the disorder are smaller and grows more slowly; tissue in the calyx area is lower in Ca and B, higher in polyphenol oxidase activity and total soluble solids content.

c) Carbon Dioxide Injury

Pears are considerably less tolerant of CO_2 injury than apples. Injury is first evident as browning of the interior carpel walls. As the disorder progresses, the core tissue adjacent to the carpels turn brown. More extensive injury of cortical tissue turns the tissue light brown in unriped fruit. As desiccation proceeds, cavities develop that may be either small and scattered or large enough for a depression on the surface to occur. Skin and sub-epidermal tissue are rarely affected.

d) Chemical Injury

During handling and storage, chemical injury occurs in fruits due to exposure to chemicals that could be injurious through misuse or because of fruit sensitivity. Chemical includes use of scald inhibiting chemicals, fungicides, calcium salts to control breakdown and bitter pit, salts to increase fruit buoyancy in flotation systems, detergents to clean fruit and waxes to enhance appearance.

e) Freezing Injury

Long exposure to mild freezing temperature results in a layer of healthy tissue immediately below the skin, followed by a layer of water-soaked tissue. The cortical tissue in such fruit shows brown vascular bundles, with brown cortical tissue in more severe cases. Severe freezing causes cavitation in the cortex.

f) Core Breakdown

Core breakdown is a senescent disorder in which core tissue becomes brown, watery and, easily separated from the healthy tissue. Main vascular elements also turn dark brown. Factors that influence core breakdown are crop load, harvest date, storage temperature, delayed cooling and delayed storage.

Management

Controlled Atmosphere (CA) storage extends the storage life of pears and thus, reduce the development of core breakdown.

g) Pink end

It is also known as premature ripening. Early yellowing of fruit at the calyx basin, accelerated softening and pink colouring of the fleshy calyx (pink end) indicate premature ripening. Serious losses can occur unless the fruit is harvested at the first signs of ripening, cooled quickly and sold promptly. Depending on the severity of the disorder, harvested fruits may continue to ripen and undergo complete breakdown, if the fruit is placed in warm temperatures after cold storage. Breakdown associated with premature ripening is somewhat different from core or senescent breakdown. It occurs at the calyx basin, where the soft light brown colour of affected tissue is clearly visible through the skin of the fruit, in contrast to core breakdown, which is not visible externally.

Pink end can be initiated, when night temperatures of about 7°C or lower occur during the last 4 weeks of fruit growth. When day temperatures are 20°C or lower, the accumulation of about 25 hours of night temperatures of 7°C or lower is sufficient to stimulate ethylene production and the onset of fruit ripening.

Management

Gibberellic acid applied about 4 weeks before harvest counteracts the effect of low temperature.

h) Senescent Scald

Dark brown skin discoloration begins as small spots and develops into large areas of skin turning brown after long term storage; brown skin may slough off. Prompt cooling and maintenance of optimum fruit temperature minimize this disorder.

i) Superficial Scald

Diffuse brown skin discoloration occurs after several months of storage and during ripening after storage. Scald inhibitors, such as ethoxyquin, may be used on pears that are stored longer than 3 months. Controlled Atmosphere (CA) storage delays scald incidence and severity.

❑❑❑

23

Peach

a) Split Pit

Split pit peaches are flattened at end of the fruit. Often symptoms are not seen until the fruit is cut open. The pit is torn apart or split along the dorsal-ventral suture. More fertilizer and water have more fruit with split pit.

Management

- Avoid excessive nitrogen applications, especially close to harvest time.
- Use stricter packing house grading for problem varieties to remove questionable or misshapen frit. These fruit are most at risk for split pits.

b) Chilling Injury

Peaches ripen and deteriorate quickly at ambient temperature. Therefore, cold storage is used to slow these processes and decay development. However, chilling injury (CI) limits the storage life of peaches under low temperature. It has been widely reported that the expression of CI symptoms, especially flesh browning or internal browning, develops faster and more intensely, when susceptible fruits are stored at temperatures between 2.2 and 7.6°C (killing temperature zone) than those stored at 0°C or below but above their freezing point.

Chilling injury visual symptoms develop within 1 or 2 weeks, when fruits are stored at 2-5°C. Chilling injury is genetically influenced and triggered by a combination of storage temperature and storage period. It manifests itself as dry, mealy, woolly (lack of juice) or hard textured fruit with no juice (leatheriness), flesh or pit cavity browning and flesh bleeding or internal reddening. In the more advanced stages, chilling injured fruit also show flesh tissue separation and cavity formation. This is more frequently observed in white flesh peach cultivars.

Management

- Foliar sprays of calcium on peach (applied every 14 days, starting 2 weeks after full bloom and continuing until 1 week before harvest) prevented the onset of Chilling injury symptoms.
- Moderate stress during vegetative growth.
- The use of more efficient training systems which allows for more sunlight penetration into the center and lower canopy areas is recommended to reduce the number of shaded fruit, thus extending post-harvest life and reduces chilling injury.
- Preharvest application of gibberellic acid reduces the chilling injury.
- The controlled atmosphere storage of peaches have found that lowering O_2 and raising CO_2 in the storage atmosphere conferred benefit on the fruit and delayed or prevented the appearance of mealiness, internal reddening and flesh browning.

c) Black Staining

It is a physiological problem affecting only the skin of peaches which is characterized by black or brown spots or stripes. These symptoms appear generally 24-48 hours after harvest. It occurs as a result of abrasion damage along contamination with heavy metals, such as iron, copper and aluminium. This occurs usually during the harvesting and hauling operations, although it may occur in other steps during post-harvest handling.

Management

- Reduce fruit abrasion damage.
- Treat fruit gently.
- Avoid dust contamination on fruits.
- Check water quality for contamination with metals (Fe, Cu & Al).

- Do not spray foliar nutrients containing Fe or Cu during fruit maturation.
- Stop pre-harvest applications of the fungicides and foliar nutrients within 15 days before harvest.

d) Blackening of Peach

The mesocarp tissue of ripening peaches becomes greyish black, when the temperature exceeds 39°C for a few days. The discolouration is difficult to detect in yellow-fleshed cultivars but is readily visible in white-fleshed cultivars. The darkening is worse near the pit than in the epidermis and sub-epidermal cells of the mesocarp, suggesting that the oxygen supply is insufficient adjacent to the endocarp. Apparently, the distance that oxygen must diffuse from the external atmosphere to the inner zones of the mesocarp is too great and the rate of movement too slow to replenish the oxygen used for respiration by the over heated cells.

24

Strawberry

a) Sunscald

Sunscald is caused by high temperatures and over-exposure of strawberry fruit to the sun. Symptoms are more common, when hot sunny weather occurs after a prolonged period of cool cloudy weather. Symptoms appear on the upper side of the fruit, near the calyx, just before ripening. The fruit becomes pale and mushy at first, eventually drying down to a distinct, firm, bleached lesion ranging in colour from pink to off-grey. Berries growing on the south or west side of rows and those without adequate foliage canopy, are most susceptible. Some varieties are more susceptible than others.

b) Uneven Ripening

This condition is thought to develop, when high temperatures occur during the fruit ripening stage affecting the enzymes that cause normal pigmentation in ripe fruit. Fruits which are exposed to high temperatures earlier in their development are not as susceptible as fruits exposed to early on cool temperatures.

c) Poor Pollination

Poor pollination can cause poor fruit set, low yields and poorly developed or misshapen fruit. The fruit will usually exhibit smaller seeds in the deformed area compared to the surrounding healthy area of the fruit. Poor weather conditions during bloom increase the likelihood of poor pollination. Powdery mildew can also cause small fruit with small undeveloped seeds. Diseases, such as powdery mildew can infect pollen and also lead to symptoms of poor pollination. Growers should take care when using insecticides harmful to bees, especially during bloom.

d) Albinism

Albinism (lack of fruit colour during ripening) is a physiological disorder in strawberry. It is probably caused by certain climatic conditions and extremes in nutrition. Fruits remain irregularly pink or even totally white and sometimes swollen. They have acid taste and become less firm. Albino fruits are often damaged during harvesting and are susceptible to *botrytisalli* infection and decay during storage.

❑❑❑

25

Plum

a) Internal Browning

It is a physiological disorder associated with high temperatures during fruit maturation and delayed harvest. It causes flesh browning, flesh mealiness, black pit cavity, flesh translucency, red pigment accumulation (bleeding) and low flavour.

Management

- Avoid cultivars susceptible to internal browning
- Market susceptible cultivars according to their potential post-harvest life
- Segregate fruits according to their postharvest potential life.
- Pick fruit well-matured.
- Enforce proper post-harvest handling during transport and at the retailer.
    - Keep fruits near 32°F during storage and transportation.
    - Avoid 36-46°F temperatures during retail handling.
- Educate warehouse and retail managers on internal browning.

Heat spot

High temperatures may cause purplish spots on plum fruits that exude clear gum. These spots develop about 5-8 days after a period of hot weather. Summer cover crops and mulches help to reduce this physiological disorder.

Management

To reduce this disorder, spray with 4 kg of calcium chloride in 1,000 litres of water during the fruit development.

❑❑❑

III. Vegetables

1

Tomato

a) Blossom end Rot

Brown water soaked discolouration appears at the blossom end of the fruit, where the senescent petals are attached while the fruit is still green. The spots enlarge and darken rapidly and the affected portion of the fruit becomes sunken, leathery and dark coloured. This disorder may occur due to:

- Sudden change in the rate of transpiration specially in moisture stress condition.
- Continuously high evapo-transpiration regime and a large leaf area.
- Increasing level of nitrogen content in the fruits.
- Deficiency of calcium.

Management

- Increase the frequency of irrigation.
- Judicial fertilizer application.
- Increase the level of phosphorus fertilizer application.
- Liming decreases the incidence.
- Foliar spray of 0.5% calcium chloride ($CaCl_2$) at the time of fruit development.

b) Fruit Cracking

Cracking of tomatoes is very common. Four types of cracking - radial, concentric, burst and cuticular are observed. Concentric and radial are very severe. In radial cracking, surface of the full ripe fruits cracks radially from the stem end of the fruit and in concentric cracking, surface of mostly mature green fruits cracks concentrically around the shoulder of the fruit. Radial cracking is more common and causes greater loss than concentric cracking. Cracking may occur due to:

- Irrigation or rainfall after long dry spell.
- Exposure of fruits to sun due to pruning and staking.
- Boron deficiency
- Genetic factor which is reported to be inherited polygenically.

Management

- Irrigation at regular intervals.
- Picking of fruits before fully ripe stage which reduces radial cracking.
- Spraying on seedlings before transplanting with 0.3-0.4% boric acid solution and repeating the sprays at the time of fruiting, 15 days after second spray and at fruit ripening stage.

c) Sun Scald

Exposed portion of either green or nearly ripe fruits get blistered and water soaked due to extreme heat of scorching sunshine. Rapid desiccation in the blistered portion turns the sunken areas white or grey in green fruits and yellow in pink or red fruits. Sun scalding may cause inhibition of fruits softening, production of tough tissues and differential fruit ripening. This problem is common in plants that have lost considerable foliage. Cultivars with heavy foliage cover are least affected by this disorder.

Management

- Protection of plants from defoliation by diseases and insect pests.

d) Puffiness

Outer wall of the grown up fruits (two-third normal size) continues to develop normally but growth of the remaining internal tissues (Placenta, mesocarp) is retarded resulting in partially filled fruit which is light in weight and lacks in firmness. Cross section of the affected fruit shows emptiness or pockets. This disorder occurs due to:

- Non-fertilization of ovules
- Embryo abortion after normal fertilization
- Necrosis of vascular and placental tissue after normal development of the fruit
- High temperature and high soil moisture are the primary factors responsible for this disorder.

Management

- Over irrigation should be avoided.
- Less nitrogen should be applied.
- Boron should be applied by spraying 0.3-0.4% boric acid solution.

e) Cat Face

Distortion of the blossom end of the fruit gives rise to various ridges, furrows and indentation in a localized area of the fruit. Cat face is named due to these ridges and indentations. Abnormal growing conditions during formation of the blossoms appear to cause death of the cells of the blossom end of the ovary which turn dark to form a leathery blotch at the end of the fruit. Abnormal growing conditions, such as extreme heat, drought, low temperature, cloudy weather during formation of the blossom appear to cause distortion of growth of the cells of the pistil. As a result, the cells in the blossom end of the ovary die and turn dark to form a leathery blotch at the end of fruit without progress of symptoms characteristic of blossom-end rot.

Management

- Select suitable variety for season.
- Adjust date of sowing and planting as per climatic requirement of the crop.

f) Blotchy Ripening

Greenish yellow whitish patch on tomato, mostly on stem and portion of the fruits is called blotchy ripening. It is caused mainly due to potassium deficiency, it also occurs due to imbalance of nitrogen and potassium.

Management

To control blotchy ripening, potassium should be applied adequately and temperature should be controlled. Short photoperiod and relatively low day temperature reduces it's incidence.

g) Gold Fleck

In the fruit surface around the calyx and fruit shoulder, tiny yellow spots often appear which are called gold flecks. These gold flecks appear due to deposition of calcium oxalate. With high incidence, fruits become non-attractive and their shelf life also gets reduced. Incidence of this disorder is increased due to:

- Higher supply of phosphorus fertilizers.
- Higher supply of calcium fertilizers.
- Increased magnesium concentration in the fruits.
- Providing shading during summer.

h) Unfruitfulness

Temperature, particularly a night temperature has profound effect on fruit setting in tomato. High day temperature (above 32°C) and high night temperature (above 20°C) is not congenial to fruit set. On the other hand, fruits normally fail to set at or below 13°C. Both high and low temperatures adversely affect fruit set mainly by reducing pollen viability and pollen germination on stigma. For this reason, low fruit set is a problem during summer cultivation in eastern India.

Management

- Growing high temperature tolerant varieties
- Application of growth substance, like parachlorophenoxy acetic acid (PCPA) @ 50 ppm at full bloom stage or 2,4 dichlorophenoxy acetic acid (2,4-D) @ 1-2 ppm prior to anthesis of the first flower cluster and again 15 days after anthesis helps to set fruits during summer and winter seasons.

i) Green Back

This disorder is characterized with greening of stem end portion. It is mainly due to high temperature and defoliation.

Management:

Reduction in the incidence of Green Back has been found by supplying potassium to plant and by shading of tomato plant.

j) Silvering

In this disorder, there is appearance of pale silvery green patches on leaves, stems and fruits. The main shoots are more affected than side shoots. The upper leaves turn silvery and the tips of the plant are severely deformed. This disorder is intensified with the increase in solar radiation and air temperature and decrease in air humidity. It is reported that silvering symptoms of tomato are due to the genetic effect.

Management

Silvering can be eliminated by utilizing cyto-plasmic variant.

k) Bronzing and Gray-Wall

There are two types of bronzing - one is associated with the TMV (Tobacco mosaic virus), while another type is produced in plants grown under the condition of shade and low temperature. It occurs in green fruits and symptoms appear in ripe fruits, when associated with TMV, while another type have been characterized with necrosis on the fruit wall, which is not associated with calyx end is less common. Gray-wall is characterized with the symptom of irregular blanched areas, mostly on the side or near stem end. It is due to low light intensity.

l) Stem Abnormality

This disorder is observed, when tomato is planted late in summer. It is characterized with hollow stem and slanky growth. It is attributed to factors like high nitrogen, low potassium, drought, soil alkalinity, low boron content in apical buds, etc.

Management

It can be managed by judicious use of nutrient as well as by following improved package of cultural practices.

m) 2, 4-D Injury

Tomato is very sensitive to 2,4-D. The symptom of this injury is the downward curvature of the leaves and tips of the growing point. Stem becomes thick, stiff and brittle with whitening of external surface. The fruits turn pointed at the blossom end and show malformation.

Management

It is advised not to spray 2,4-D in the neighbouring field, when there is a high wind velocity towards the tomato field.

n) Dry Set and Chat Fruit

This problem arises due to low temperature and poor light condition. In this environment, pollination and fertilization process gets inhibited and this is considered as dry set and if these fruits develop up to 30 mm and these are called 'Chat fruit', when ripe.

Management

It can be controlled by manipulating the environmental condition.

o) Rain Check

Rain check can be described as tiny cracks that develop on the shoulder of the fruit. These cracks can vary from just a few to almost complete coverage of the shoulder. The cracks feel rough to the touch and affected areas can take on a leathery appearance and not develop proper colour as the fruit ripens. Green fruits are most susceptible followed by breakers and are not affected at all. Damage occurs most often on exposed fruits after a rain. The problem is more severe, when heavy rains occur after a long dry period. There are differences among varieties to susceptibility to rain check. Also, varieties with good leaf coverage usually have less rain check.

p) Parthenocarpy

Parthenocarpy is the ability to develop fruit in the absence of fertilization, is a desirable genetic trait in vegetable crops which are grown for commercial value. The advantages conferred by Parthenocarpy are fruit production under environmental condition prohibitive for fertilization and production of seedless fruit. Parthenocarpy has been often linked to aberrant sexual development. In several tomato mutants and transgenic lines defective in stamen, anther and ovule development have been described to produce parthenocarpic fruit.

There are two types of Parthenocarpy viz., Natural Parthenocarpy and Artificial Parthenocarpy.

(i) Natural Parthenocarpy

Natural occurring Parthenocarpy has been identified as obligate and facultative.

- **Obligate Parthenocarpy:** Genotypes with obligate Parthenocarpy are unable to produce viable seeds either in the presence or absence of fertile pollen. Plants with this condition are thought to be defective during female gametophyte development and require manual vegetative propagation.
- **Facultative Parthenocarpy :** Facultative parthenocarpic plants produce seedless fruits only when pollination is prevented by growing conditions (e.g. temperature), or by manual manipulations (e.g. emasculation) or physiological barriers (e.g. incompatibility).

(ii) Artificial Parthenocarpy

Parthenocarpy can be artificially induced by external application of plant growth hormone or by altering the endogenous plant hormone levels within the ovules or carpel tissue. Parthenocarpy can be induced artificially by following methods:

- External application of plant growth regulators (PGRs)
- Molecular or transgenic approach.

Causes for Parthenocarpy

- Aborted Embryo.
- Chromosome Imbalance.
- Self-Incompatibility.
- Growth Regulators.

Advantages

- Seedless tomato fruits exceed seeded fruits in dry matter content by upto 1%, more sugars and more soluble solids.
- The shelf life of seedless fruits is longer.
- Parthenocarpic tomato plants produce a high yield and fruit set in colder temperature than seeded cultivars.
- Studies have shown that seedless tomato fruits are tastier than the seeded variety.
- Parthenocarpic plants often display an early fruit set, taking place before anthesis that allows early fruit production.

q) Internal White Tissue

Internal white tissue is a disorder which shows no outward symptoms but when ripe tomato is cut, it reveals white hard areas especially vascular region of the outer walls. Under severe conditions, the fruits may show white tissue in the centre of the fruit. This can make fruits unmarketable. High temperature during ripening period in the field triggers the development of white tissue. Adequate Potassium application has shown to reduce the development of white tissue.

r) Physiological Leaf roll

The leaf roll symptoms are as follows: Upward cupping of the leaf margins followed by an inward leaf roll so that the margins of adjacent leaflets may overlap in a tube-like appearance. Lower leaves are affected first and rolling may eventually involve most of the plant. Rolled leaves feel rough and leathery to touch but are otherwise normal in size and colour. Leaf roll symptoms typically persist through the season, particularly if conditions favouring leaf roll are prolonged.

Rolled leaves can sometimes recover if environmental conditions and cultural factors are adjusted to reduce plant stress. Plant growth, fruit yield and quality do not appear to be affected.

(i) Physiological leaf roll can be confused with

Symptoms expressed following some virus infections, in addition to leaf curl, often include yellow chlorosis, mottling and stunting of leaves. Growth regulating chemicals damage (herbicide application, herbicide-contaminated mulch or manure, or saliva of pests like aphids). Symptoms depend on exposure and age of plant tissue at the time of contact and may involve pointed, narrow leaves, prominent veins, young leaves that don't expand, as well as fruit deformities.

(ii) Factors affecting physiological leaf roll

- The severity of leaf roll appears to be cultivar dependent. High-yielding cultivars tend to be most susceptible. Indeterminate-growth ("staking"/ "vining") tomato cultivars appear to be more sensitive to this disorder than determinate growth ("bush") types.
- In some cases, the condition occurs most commonly when plants are staked and pruned under dry soil conditions.
- Nutrient stress, including growing high yiedling cultivars under abundant nitrogen content; or plants experiencing phosphate deficiency.
- Drought or extended dry periods.
- Excessive soil moisture under prolonged high temperatures can induce leaf roll.
- Transplant shock may result in leaf roll symptom development.

Management

- Careful cultivar selection to avoid susceptible cultivars.
- Plant in well-drained soils and maintain uniform, adequate soil moisture.

❑❑❑

2

Chilli

a) Fruit or Flower Drop

The factors responsible for this problem are extreme high temperature, low light intensity, shade, water deficit stress, biotic stress and presence of rapidly growing fruits on plant. The most serious factor responsible for this problem is high temperature; whereas high night temperature is rather more detrimental to fruit set than day temperature. Abscission of flowers becomes more serious under high temperature and water deficit conditions. Water deficit at normal temperature may not be enough to abscise the flowers. Low light intensities and shade, which are general features of glasshouse, result in poor fruit set in. Excessive use of nitrogen disturbs the C:N ratio, which also results in poor fruit-set. Flower dropping is due to the formation of abscission layer at the base of pedicel.

Management

- It is a common problem in chilli cultivation.
- Spray any one of the plant growth regulators viz., NAA (Naphthalene Acetic Acid) @ 20-40 ppm (20-40 mg/litre), first at bud initiation and second spray at 20 days later, or
- 2-4-D (Sodium salt) @ 2-4 ppm (2-4 mg/litre of water) applied at foliar spray at flower bud initiation and 20 days later.
- Both will increase yield upto 30-40 per cent over unsprayed plots.

b) Abnormal Fruit Shape

Faulty pollination and fertilization cause abnormal shape of fruits. The faulty pollination may be due to injury caused by insect, or low or high temperature to the stigmatic portion. On one side ovules get fertilized, but other side not. Seeds in fruits are the sources of auxin synthesis, which is essential for the enlargement of ovary cells. As a result, on one side where the seed content is high, the fruit growth is more and on other side where the seed content is less, the fruit growth is poor. This difference in growth makes the fruit shape abnormal. This condition generally occurs more, when the crop is grown at a temperature lower than optimum required for normal pollination and fertilization. The fruits with less seed content are usually smaller with thinner pericarp and of flat irregular shape than normal ones. The shape of fruit is largely influenced by temperatures prevalent during ovary development or by absence of seeds.

c) Blossom end Rot

Blossom end rot is caused by a localized calcium deficiency in the fruits. With water, the plants absorb not only calcium but absorb several other nutrients also and water absorption is associated with high rate of transpiration. Hence, the organs having highest rate of transpiration accumulate more calcium. Even in normal situations, the calcium level in the fruit remains very low. Excessive moisture in the soil makes the calcium unavailable to the plant. On the other hand, water deficit condition also reduces the calcium uptake by the plant. Due to the deficiency of this element, the blossom end portion becomes sunken, brown to black in colour, thick and leathery. Large fruited varieties are more susceptible to this disorder.

d) Sunscald

Sunscald is the after effect of extreme sunlight. The fruits after attaining mature green stage if exposed to high intensity light become susceptible to tissue damage. Bleaching of sun-exposed portion is called sunscald. Initially, this portion becomes silvery white due to chlorophyll disappearance and later, it is converted to black colour. This injury is developed by a combination of high temperature and high light intensities. Bell pepper has been found more susceptible to this disorder than hot pepper. This disorder is generally prevalent in regions with high light intensities and high temperature.

❑❑❑

3

Brinjal

a) Blossom end Rot

Blossom end rot (BER) also affects eggplant, which usually begins as a small water-soaked area at the blossom-end of the fruit that subsequently develops into to a dry rot. However, eggplant is appreciably less susceptible to BER than tomato. This disorder is due to the deficiency of calcium.

b) Calyx Withering

The symptoms of this disorder become apparent seven days before harvest. The affected fruits become reddish brown in colour and lacking in normal luster and thus marketability of the fruits is hampered. The affected fruits have much higher calcium and nitrate contents than healthy ones.

c) Poor Fruit Set

In brinjal, four types of flowers, according to length of style, are recognized irrespective of the varieties namely, long-styled, medium-styled, pseudoshort-styled and short-styled. Pseudo short and short-styled flowers do not normally set fruits but their numbers in a plant are normally higher than long-and medium-styled flowers which produce fruits. In brinjal, flowers are normally borne solitarily as well as in clusters. The solitary flowers are mostly long-or medium-styled; whereas mixtures of short-styled, pseudoshort-styled and medium/ long-styled flowers are found in the cluster. Obviously, ratio of short-styled flowers to long/medium-styled flowers in the cluster is much high (3-4: 1). So, fruit set depends on the number of productive flowers (long/medium-styled) in the plant. Natural drop is also a problem in brinjal which may be as high as 50% in medium-styled flower and 30% in long-styled flower. Preponderance of unproductive flower types and flower drop sometimes result in poor fruit set in brinjal.

Management

True short-styled flowers are actually sterile flowers which cannot be induced to set fruits. However, application of growth substances on pseudo short-styled, medium-styled and long-styled flowers produce higher percentage of fruit set.

- Spraying the plants with 2, 4-D @ 2 ppm at flowering stage, when few flower clusters appear.
- Spraying the plants with 60 ppm NAA (Naphthalene acetic acid) or 50 ppm PCPA (Parachlorophenoxy acetic acid) at full bloom stage. Application of growth substances also produce earlier and more uniform fruits.

d) Internal Fruit Rot

Internal fruit rot is another physiological disorder that affects egg plant fruits. Internal fruit rot seems to be a calcium-related physiological disorder. The early symptom of this disorder is the appearance of an uneven wet area on the external fruit surface, which is localized mainly in the distal region. Below this uneven area, the fruit tissue appears wet, blackened and disintegrated, presumably as a result of cell bursting and disorganization. The dark colour of the disintegrated tissue apparently originates from the disorganized cells. At a later stage, the fruit skin may crack and black juice may be excreted.

e) Chilling Injury

Egg plant fruits are chilling sensitive at temperatures below 10°C (50°F). At 5°C (41°F), chilling injury will occur in 6-8 days. Consequences of chilling injury are pitting, surface bronzing and browning of seeds and pulp tissue. Accelerated decay by *Alternaria* spp. is common in chilling stressed fruit.

❑❑❑

4

Onion

a) Bolting

It is a serious disorder of onion. It refers to the emergence of seed stalk prior to time of their formation and adversely affects the formation and development of bulbs. Bolting is an undesirable character because, it directly affects the bulb yield of onion. This disorder is due to so many factors.

- Rabi onion starts to bolt at 20-25°C temperature. Early transplanting (August) and late (end of December-January) induce bolting in kharif and Rabi onion respectively. The restricted or poor vegetative growth also leads to bolting.
- Low temperature (10-12°C).
- Varietal difference-white cultivars are more sensitive to bolting.
- Late transplanting of seedlings.
- Transplanting of aged seedlings of above 10 weeks.
- Poor supply of nitrogen in nursery and field.

Management

- Adjust the time of transplanting in such a way that the crops may be exposed to moderate temperature at bulbing. Maturity of rabi crop coincides with high temperature compared to kharif crop.
- Grow non-bolting cultivars
- Transplant healthy and six to seven week old seedlings.
- Supply recommended dose of nitrogen.
- Cut the seed stalk at early stage.

b) Sunscald

During warm dry season, a hot sun may at times raise the temperature at the soil surface to 65°C. This is particularly the case on dark soils, such as organic ones. The heat damages the sensitive young plants, killing the cells at the neck level. The injured tissues shrivel, strangle the neck and the plant wilts and withers. The seriousness of the damage depends on the temperature at the soil surface and the tenderness of the plant tissues.

Management

Sun scald is a minor problem rarely encountered in onion crops. Plant seeds as early as possible so that the plants soon outgrow the sensitive stage.

c) Translucent Scales

This physiological disorder of the onion is characterized by greyish, watery tissues of one or more layers or scales, which make them, appear translucent. Sometimes, the condition affects all the layers, but most often, only the second and third fleshy scales. In cross section, the affected scales are brownish. Translucent scales can be confused with frost injury. Frost also produces areas of opaque white tissues within the translucent tissues. Finally, with frost injury, the skin is loose on the concave side of the affected layers and the tissues take on a granular texture. The problem seems to be connected with storage, because it appears after harvest and worsens after 3 or 4 months of storage. Research has shown that onions kept at 5-10°C for few weeks before final storage at 0°C are more subject to this disorder. Excessive relative humidity could also predispose onions to this problem.

Management

- Dry onion stored as soon as possible at a proper temperature (0°C) and at a relative humidity of 65-70%.

d) Thick-Necks

Green leaves continue to appear from the top of the partially formed bulb, the neck does not soften and the bulb does not become dormant. This problem is most severe in seasons, locations and cultivars where crop maturity as shown by foliage fall-down is already late. As might be expected from the effects of temperature on the rate of bulbing, late maturity occurs in cool summers. Late maturity is also a feature of sites where low soil temperatures result in slow emergence and seedling growth and the later development of a high LAI. High-yielding, spring-sown cultivars adapted to cool regions are probably prone to this problem because, in order to make sufficient LAI for a high yield, they must continue leaf growth long after the summer solstice. The additional stimulus for bulbing associated with a high LAI may itself be important in pushing these crops to ripeness. Therefore, the crop can be predisposed to thick-necking by any factor which destroys leaves of slows leaf growth.

e) Premature Bulbing

Bulbing very rapidly, soon after sowing, so that the plants remain dormant through much of the growing season, sometimes occurs with late summer sowings of over wintered crops. Here, seedlings emerge while temperatures and photoperiods still favour bulbing. In tropical countries, onions are normally raised at high plant densities in nursery beds prior to transplanting. High plant densities may contribute to premature bulbing via the red: far red effect.

f) Inflorescence Suppression

Failure of flower emergence is a familiar problem to breeders and often results from the suppression of fully developed inflorescences by bulbing during the 'competition' phase. This is avoided using the sequence of growing conditions that favour bolting. In the case of bulbs, a long duration near optimum temperatures for vernalization will favour subsequent inflorescence emergence. Rapid increases in temperature and photoperiod can cause inflorescence suppression.

g) Watery Scales

Watery scale is a disorder of stored bulbs, the symptoms of which are a thick leathery outer skin, which when peeled away reveals watery, glassy storage scales. The scales may be later infected by bacteria or fungi, but these are not the primary cause of the disorder. Up to 15% losses of stored bulbs have been reported, particularly in crops given high temperature drying treatments following harvest.

h) Wind and Pelting Rain Injury

Wind and pelting rain injury is caused mainly during storms, when rain drops driven by violent winds strike and injure the onion leaves. Whitish to yellowish spots, round to irregular in shape and varying in size from 1 to 5 mm, are produced. Rain injury is generally not serious, but it is important to distinguish it from damage caused by *Botrytis* spp., ozone, or certain herbicides.

i) Herbicide Injury

Improper use of herbicides can cause temporary or permanent injury to onions. Damage can occur, when herbicides are applied at excessive rates or at an ill-considered stage of onion growth. With contact herbicides, burns and necrotic spots are produced and the leaf tips wither. On the other hand, with translocation herbicides, the onions may turn yellow or the leaves may curl. Generally, the plants recover and the symptoms disappear over time. However, if the application rate is too high, the damage will be irreversible. Damage may also be observed if onions are grown where the previous year's crop was treated with a residual herbicide to which onions are susceptible.

Management

- To avoid injury, follow recommendations concerning the use of herbicides on onions.
- Know which herbicides were used on the crops preceding the onion crop.

j) Bulb Splitting

The first symptom observed is the splitting of the basal plate. Secondary growth of the affected bulb often occurs as one to several small bulbs protruding from the split basal plate. Uneven irrigation of onion fields increases the incidence of this disorder. Fields that are over-irrigated, allowed to dry completely and then over-irrigated again often have many split bulbs. These openings can provide an entrance for secondary microorganisms, which cause bulb decay. Bulb mites are frequently associated with bulb splitting. However, they have yet to be implicated as the causal agent.

Management

Attention to seed bed preparation and the use of high quality seeds will result in uniform stands, which will reduce this disorder. The maintenance of uniform irrigation and fertilization practices to prevent phases of rapid and slow growth of onion bulbs can reduce the incidence of this disease

k) Freeze Damage

Affected seedlings become yellow at or near the soil line, when temperatures are below freezing for prolonged periods. Upon freezing and thawing, the soft tissues lose their integrity and become translucent and watery in appearance and texture. Freeze damaged scales become greyish yellow colour. Often, individual scales are injured entirely but adjacent inner and outer scales may escape damage. However, the bulbs may still be unmarketable. Freezing of onions become a problem at temperatures below -20°C (280F). When soil is repeatedly frozen and thawed, the plants can be heaved to the surface of the soil where they die from root damage and desiccation. Bulbs in the ground are less likely to be freeze damaged than those on the soil surface.

Management

Onion bulbs vary greatly in their ability to tolerate freezing temperature. Onions least tolerant to freezing are usually those lowest in soluble solids.

l) Greening

Sunlight causes the formation of chlorophyll in the outer scales, which results in the scales turning green. Excessive or late season nitrogen applications can delay maturity and enhance the greening of onion bulbs. Greening can occur if onion bulbs are exposed to sunlight during the growing season or the bulbs are allowed to cure for extended periods under moderate light.

Management

An early fertilizer application that promotes foliar development can reduce losses from greening at bulb maturity. Avoid excessive and late season nitrogen application. Do not cure bulbs for extended periods in the field.

m) Leaf Variegation (Chimera)

Leaf tissue is variegated resulting in tissues that have a normal green colour being directly adjacent to tissues that are varying shades of yellow to white in colour. Variegated tissue patterns may be mosaic or linear. The yellow to white tissue is deficient in chlorophyll and can result in abnormal or stunted plant growth, when severe. This is a genetic abnormality and its expression and occurrence is unaffected by environmental conditions. This condition generally occurs on only a small percentage of the plants in a field.

Management

Plant seeds that are known to be free of genetic abnormalities.

n) Storm Damage

Injury from wind, pelting rain or hail is usually visible only on the side of leaves that were facing prevailing winds during a storm. Spots develop that are typically 1-5 mm (0.06-0.2 in.) diameter, white to yellow in colour and round or irregular in shape. Although rain damage is rarely serious, hail damage may defoliate a crop. Storm damage may be confused with *botrytis* leaf blight or herbicide injury and can also make plants more susceptible to fungal and bacterial pathogens. Raindrops, hailstones and soil particles blown by strong winds wound leaf, neck and flower stalk tissues.

Management

In areas that are subject to storms, plant cereal crops with onions to act as a wind break. Once the onion seedlings are established, use selective herbicides to kill the cereal crop. Spray broad spectrum fungicides on storm damaged plants to reduce the risk of secondary infections.

❑❑❑

5

Garlic

a) Sprouting

Sprouting of bulbs in the field is noticed sometimes towards the start of maturity stage of bulbs particularly when there are winter rains or excessive soil moisture and nitrogen supply. This disorder is, however, not of permanent in nature and varies from variety-to-variety. Early planting also causes sprouting.

b) Rubberification

The rubberification and premature sprouting of bulbs are main physiological disorders in garlic. The reasons for these disorders are furnished below :

- Rubberification and premature sprouting of bulbs are noticed mostly in fields which are located in low-lying areas of watershed where there is heavy deposition of nutrients along with slit during heavy rains.
- These problems are severe in garlic fields which are more frequently irrigated than the normal requirements of garlic bulbs.
- With the application of higher levels of nitrogen, there is an increased level of pre-mature sprouting of bulbs which results in splitting and rubberification of bulbs.
- Rubberification incidence also increases, when higher level of nitrogen is applied in the form of urea.
- Due to increased levels of nitrogen, there is an increased level of thrips incidence in leaves during the later stage of crop. The thrips lacerate leaves and cause severe damage to the crop. Through lacerated wounds by thrips, there is a severe incidence of blast disease, caused by *Botrytis allii*, when the crop is at 70-90 days

old, which further deteriorates the quality of garlic bulbs.

- Short duration type of garlic is more susceptible than long duration type.
- Premature sprouting of bulbs is more in the crop (April-August), when there is high moisture level in soil at maturity of crop due to heavy rains.
- Delayed harvest during rainy season has increased premature sprouting and splitting of bulbs.
- Wider spacing of cloves at the time of planting increases uptake of nitrogen and water by the individual plants which increases premature sprouting and rubberification of garlic bulbs.

c) Waxy Breakdown

A physiological disorder that affects garlic during the later stages of growth and often associated with periods of high temperature near harvest. Early symptoms are small, light yellow areas in the clove flesh that darken to yellow or amber with time. Finally, the clove is translucent, sticky and waxy, but the outer dry skins are not usually affected. Waxy breakdown is commonly found in stored and shipped garlic but rarely in the field. Low oxygen levels and inadequate ventilation during handling and storage may also contribute to development of waxy breakdown.

❑❑❑

6

Potato

a) Hollow Heart

It is caused by rapid growth of tubers. Tubers become oversized and remain empty within leading to the formation of cavity in the centre with the death of the small area of pith cells. This results in adjacent cracks and hollowness, as the centre expands during the growth of the potato.

Management

- Maintain soil moisture conditions to the optimum level. Avoid over fertilization particularly nitrogen.
- Grow those varieties which are less prone to this defect.

b) Black Heart

The potato is affected by a number of diseases marked by discoloration of the interior tissues of the tuber, but one of these internal necroses which causes a characteristic blackening of the center and sometimes of more external parts has been called the blackheart or, less frequently, heart rot. Internal discolorations of the tuber may be caused by either parasitic or non-parasitic influences.

The appearance of tubers affected by blackheart is exceedingly variable, depending upon the exact conditions under which the disorder has developed. In one form of the disorder, the tubers may appear perfectly normal as far as the external characters are concerned, but when cut into two, they are found to show a browning or blackening of the interior. The discolouration generally starts at the center and progresses towards the outside, causing a star-like radiation.

If the inciting conditions operate for a sufficiently long period, the blackening may advance until it reaches the surface, the blackened tissues are in sharp contrast to the normal flesh, similar in consistency to normal tissues or more firm or slightly leathery if they have partially dried. This character should distinguish the lesions from those either leak or black rot, which are similar in colour but soft and watery. In some cases of blackheart, the discolouration appears in the form of zones surrounding the center, which is normal or only slightly discoloured.

The extent and the character of the blackened tissue will serve to separate this phase of the disorder from hollow heart, in which a growth cavity is surrounded by a narrow zone of brown, oxidized tissue. If these blackheart lesions advance until they reach wounds, rot-producing organisms are likely to enter and continue the work of destruction.

Management

- Provide proper ventilation. Keep potato tubers in layers. Do not store tubers in the heap.

c) Greening

There are various factors which increase the glycol alkaloid contents such as mechanical injury, premature harvest and excessive application of fertilizers or exposure of tubers to sunlight which leads to solanin production which is slightly poisonous.

Management

- Proper earthing up of tubers as the tuberization takes place. Store tubers in darkness after digging up.

d) Knobbiness

It occurs due to uneven growth of tuber cells/tissues. Uneven watering conditions lead to an obstruction in tuber growth. Heavy irrigation after a long dry spell leads to growth of some cells very fast resulting in knobbiness.

Management

- Frequent and optimum irrigation supply is the remedy.

e) Cracking

It is due to boron deficiency or uneven water supply.

Management

- Application of Borax @ 20kg/ha. Frequent and optimum irrigation supply is the remedy.

f) Sun Scalding

It occurs, generally, in the autumn crop when the temperature is high and sunshine is more. Emergence of sprout and leaflets is drastically affected. It appears, when temperature is more than 30°C.

Management

- Water should be passed through the furrows to lower the soil temperature.

g) Translucent end

It is related to environmental stress and occurs due to drought and heat. It is generally found at the proximal end of the tuber. Tubers show glossy appearance and are irregular in shape. This also results in decay in storage. These glossy areas are high in sugar and low in total soluble solids.

Management

- Avoid excessive nitrogen supply. Maintain 50% moisture in the field.

h) Black Spot

It means the internal browning of potato tubers. It occurs in vascular tissues within three days of mechanical injury. Phenols are related to black spot in potato tubers.

Management

- Genetic makeup of the varieties. Provide proper storage and growing conditions.

i) Freezing Injury

It occurs due to the exposure of tubers to freezing temperature during or after harvest. It takes place at -1.5^{0}C or below temperature. There is discolouration of the tissues and affects the vascular tissues at ring called ring necrosis. When fine elements or cells of vascular ring are affected, it is called as net necrosis. This leads to unmarketable tubers. Tubers show more damage towards proximal end.

Management

- Avoid exposure of tubers to freezing temperature during storage or harvest.

j) Swollen Lenticels

This disorder is caused by exposure of the tuber to very wet conditions in the field or in storage due to oxygen deprivation. Besides, giving an unmarketable appearance to the tuber, the major problem is that an entrance to pathogenic organisms, bacterial soft rot and pink rot.

Management

- Avoid over-watering.
- Keep storage well-ventilated.

k) Sweetening

Holding potatoes in cold storage even at moderate temperature (10-13°C) for only few-days causes excessive accumulation of reducing sugars, which results in sweetening of potatoes and production of undesirable dark coloured processed products. During storage at low temperature, reducing sugars accumulate more swiftly than sucrose. However, the amount of sugars formed during storage at low temperature depends on variety, maturity of potatoes, pre-storage conditions, storage temperature and sprouting.

l) Dormancy

The time lag between harvest and initiation of sprouts is termed as dormancy. Dormant or even semi-dormant tubers do not sprout readily after planting causing delayed and erratic crop stand. The problem of dormancy is particularly important when :

(i) two potato crops are grown in rotation, such as, December harvesting followed by January planting which is practiced in prolonged winter condition.

(ii) tubers produced in the hills are used as seed in the plains soon after harvest.

Management

- In advanced stage of dormancy, cutting of tubers may terminate dormancy.
- Thiourea treatment: Freshly harvested clean tubers are cut into pieces, each piece having at least two eyes. The cut tuber pieces are immersed immediately in 1% thiourea solution for one hour. The treated tuber pieces should be planted immediately or left overnight in wet gunny bags for planting in the next day.
- Rindite treatment: The mixture of seven parts of ethylene chlorohydrins, three parts of ethylene dichloride and one part of carbon tetrachloride by volume is called Rindite and is applied @ 0.3 ml/kg of tubers in an airtight chamber.
- Triple treatment: Spraying the whole tubers with 3% ethylene chlorohydrins @ 20 ml/kg of tubers are dipped in 1% thiourea solution for one hour and finally in one ppm gibberellic acid solution for 10 minutes. The treated seed tuber pieces should be dried in shade before planting.

m) Sprouting

Potatoes usually do not sprout upto 2-3 months after harvest even at moderate temperature (10-15°C) because of their endo-dormancy. However, after 2-3 months of storage, sprouting can be anticipated in potatoes stored at temperature nearby 15°C. Storage at 4°C is considered optimum for most of the cultivars for maximum storage life (5-8 months) because sprout growth is absent, shrinkage is low and other losses are also less. If longer storage is required, the potatoes should be stored at 3°C temperature.

Management:

- Maleic hydrazide (MH) is applied on plants in the field, 15 days before harvest when the leaves are green.
- Isopropyl-N-3 chlorophenyl carbamate (CIPC) is extensively applied to potato after harvesting.
- A combine spray of CIPC, thiabendazole and chlorine applied to potatoes at the time of storage overcomes the detrimental effects of CIPC and also reduce *Fusarium rot.*
- HPP (hydrogen peroxide plus) can also inhibit potato sprouting during storage.

n) Tip Burn

Tip burn is the first in evidence as a slight wilting and yellowing of the tissues at the extreme tips of potato leaflets or more rarely, at the margin at some point back of the tip as well as at the tip. This yellowing is soon replaced by a browning and death of the tissues and the dead area is extended from the tip downward or from the margin inward until in extreme cases, entire leaflets are brown and dead. The browning begins at the tip or margin and gradually advances downward or inward into the pale-green or chlorotic tissue.

Under favourable conditions for the development of the disease, the lesions may gradually advance, or the progress may be delayed or checked by the return of better growing conditions. The amount of tip burn varies with the position of the leaves and is also influenced by the age and maturity of the leaves. Young leaves which are standing nearly upright appear to suffer the least, while the tip blight is more severe on those leaves which are older and occupy a position so that the sun strikes them more nearly at right angles to their surface.

The older leaves drop so that only the tips of the end leaflets are exposed, but those of all of the leaflets may hang down and be struck by the tip burn, the oldest leaves are likely to be under the shade of those from the middle of the stem and so suffer less than those in the middle. If they are exposed at all, they succumb readily and all of the leaflets die in a few days. The earlier attacks, however, are likely to occur on the middle leaves, and after they have become thoroughly scorched and dead, the oldest leaves are exposed and in turn succumb.

o) Internal Rust Spot

Internal Rust Spot is characterized by dry, corky, reddish-brown blotches in the flesh of the tuber, generally towards the 'rose end'. Symptoms are present at harvest and do not increase during storage; nevertheless, they detract significantly from tuber quality. This disorder results from a localised calcium deficiency, large or long tuber being especially prone. Cultivars differ in susceptibility to internal rust spot.

Management

An adequate water supply to the growing crop, especially in hot climates, where drought stress (and associated mineral deficiency) may be a problem during tuber development.

p) Jelly-End Rot

Jelly-end rot is physiological breakdown of the heel-end of the tuber. The affected part first appears translucent or 'glassy' and later becomes soft and jelly-like. Bacteria may cause secondary breakdown or, under dry conditions, the tissue loses moisture and becomes leathery. The disorder is one of the consequences of 'second growth' of the stems and foliage, which occurs if conditions become favourable after a period of drought, excessive temperatures or other adverse factors. The sudden spurt of growth causes withdrawal of starch from the tuber and consequent physiological breakdown.

Management

The main preventive is to avoid water stress during early stage of growth.

❑❑❑

7

Sweet Patato

a) Dry matter Loss and Pithiness

Sweet potatoes lose dry matter through natural respiration. Respiration is a chemical process necessary for all the living tissues, whereby starches and sugars (dry matter) are oxidized to carbon dioxide and water vapour with the liberation of heat. The heat generated by an individual sweet potato is negligible, but the combined output of tonnes of sweet potatoes in a storage facility can raise the temperature of sweet potatoes one-fourth to one-third degree per day. This heat must be continuously removed from the facility, or the temperature will rise above acceptable levels in a short time. A storage facility must have provisions for cooling with outside air or a refrigeration system. Although not apparent externally, significant dry matter loss may result in pithiness with the formation of many small voids. Pithiness is very common in sweet potatoes held for long periods in poorly controlled storage facilities.

b) Sprouting in Storage

Another effect of elevated storage temperatures is sprouting. At temperatures above 60°F (16°C), sweet potatoes will sprout. The length of time required for sprouting depends on the temperature. It may take a month or more for sprouts to show at 65°F (18°C), but at 75°F (24°C) and warmer, sprouts can develop in a few weeks. Sprouting is always accompanied by rapid respiration and weight loss.

c) Chilling Injury

Chilling injury is rare in modern storage facilities, but it can occur if roots are kept in common areas during the winter months. Storage below 50°F (10°C) can result in chilling injury that may not be evident until several weeks have passed.

d) Excessive Shrinkage

If the humidity is low, sweet potatoes will lose weight as moisture evaporates from the surface of roots. This results in weight loss and may cause shrivelling of the skin, especially at root ends. Although some moisture loss is practically unavoidable during curing and storage, excessive water loss may be avoided by maintaining high relative humidity during storage.

❑❑❑

8

Radish

a) Forking

There is secondary elongating growth in the roots that gives a look of fork like structure to the root. Inter varietal variations are being considered to result from the degree of secondary elongating growth. The disorder is due to the excess moisture during the root development of radish. It also occurs in heavy soil due to the soil compactness. Un-decomposed organic manure favours elongated root in radish.

Management

It can be corrected by reducing the moisture from the field by balanced irrigation and also by sowing radish in sandy loam or light soils of loose and friable in nature. Always apply well decomposed organic matter.

b) Pungency

The taste of radish for fresh eating is one of the most important quality traits. The pungent component of radish is 4-methylthio-3-butenyl isothiocynate (MTB-ITC). It was confirmed that many varieties (temperate) were low in the contents of MTB-GSL (100-200 mmol per 100 g fresh weight). It happens due to high temperature and water stress condition.

Management

Proper water management. Selection of variety as per the season.

c) Rough Roots

Radish crops grown in heavy soils produce rough, misshapen roots with a number of small fibrous laterals and such soils are to be avoided. Since the crops grow so rapidly, rich fertile light friable loamy soil can overcome this malady.

d) Scorching

If the environmental temperature is high and exposure to light is prolonged, the tuber turns bronze-green and the cells below the bleached area die.

e) Strings of Tubers

The tubers are small and in a row. This is frequent in late varieties due to several successive interruptions to tuber formation.

f) Cracks and Hollowness

These are due to changes in environmental factors, such as water and temperature and excess nitrogen availability in the later phases of growth.

g) Brown Heart

It is characterized by the appearance of dark spots on the thickest part of the roots. The growth of the plants is stunted. The leaves become smaller and lesser in number and show variegated appearance with yellow and purplish red blotches. The leaf stalks show longitudinal splitting. The roots remain small, distorted and greyish.

Management

It can be rectified by soil application of borax @ 15-25 kg/ha or foliar application of 0.1% boron.

h) Pithiness

Pores are formed in the root. The quality of radish is reduced, destroying its commercial value. Physiologically, parenchymatous cells in root tissue are collapsed. Pore development is a sign of senescence and its degree varies from variety to variety. Varieties forming early pores have larger parenchymatous cells in roots, more tender basal tissue and lower starch content. Varieties showing early maturity, earlier enlargement of root and rapid reduction in T/R ratio (ratio of the weight R to the top weight T) have a tendency to form large pores. When harvesting is delayed, this disorder is more pronounced.

Management

Harvesting should be done at appropriate time.

❑❑❑

9

Carrot

a) Wart

White inner protrusion of inner root tissue through splits in skin occurs. It is mainly due to soil-moisture deficiency.

Management

Proper water management.

b) Root Splitting

Splitting or cracking of carrot roots is a major problem in many carrot growing areas. As clear by its name, roots get splitted making it unfit for market. Earlier cultivars split more readily than late ones. In soils with high content of N, splitting of roots is increased. High soil concentrations of ammonium compounds caused more serious splitting than by other forms of nitrogen. When the roots were 6.5 cm in diameter, increased splitting occurs especially when nitrogen is applied to widely spaced roots. Wider the spacing, the greater chance of splitting. If roots are large, there are chances of splitting as compared to small ones.

Management

Follow proper management practices like right spacing and optimum amount of N application. Prefer nitrogen fertilizer source other than ammonium compounds.

c) Cavity Spot

This disorder appears as a cavity in the cortex and in most cases, the subtending epidermis collapses to form a pitted lesion. The cavity spot disorder is induced by deficiency of Ca. This is associated with an increased accumulation of K which leads to a decreased accumulation of Ca. Cavity spot disorder of carrot roots is a manifestation of Ca deficiency, which may be induced by excess K uptake during the ontogeny of carrot plants.

Management

Increase in Ca level in the growing medium leads to significant reduction in the incidence of cavity spot. Application of fertilizers should be kept minimum and only be done in spring or early summer. This would avoid a build up of K in the plant.

d) Bitterness

Bitterness is considered as an undesirable taste of carrots. Quantitative chemical analysis of potential bitter compounds of different carrot genotypes is combined with sensory analysis in order to identify key compounds likely to be responsible for the bitterness of carrots. Bitterness in carrots, which may develop in storage, is due to abnormal metabolism caused by ethylene. This gas is given off by apples, pears and certain other fruits and vegetables and from decaying tissues.

Management

Bitterness can be prevented by storing carrots away from such produce. Also, development of bitterness can largely be avoided by low-temperature storage, as it minimizes ethylene production.

❑❑❑

10

Beetroot

a) Internal Black Spot or Brown Heart

Boron deficiency may cause a physiological disorder in garden beets which is known as internal black spot or brown heart. Boron deficient plants usually remain dwarf or stunted. The affected leaves are smaller than normal ones. The young unfolding leaves fail to develop normally and eventually turn brown or black or die. The leaves may assume a variegated appearance due to development of mixture of yellow and purplish red blotches over parts or whole, while the stalk of such leaves show longitudinal splitting. Frequently, the affected plants have twisted leaves and exhibited a slight shortening and distorting of its leaf-stalk in the centre of its crown. The growing point may die and decay. The roots do not grow to full size and under conditions of severe boron deficiency, they remain very small and distorted. They may also have a rough, unhealthy and greyish appearance instead of being clean and smooth. Their surfaces are often wrinkled and cracked. Within the fleshy roots, hard or corky spots are found scattered throughout the roots but more numerous on the light coloured zones or cambium layers.

Management

The disorder can be effectively controlled by application of borax to the soil @ 5 to 15 kg per hectare.

❑❑❑

11

Turnip

a) Brown Heart

Boron deficient plants usually remain dwarf or stunted. The leaves are smaller than normal. The roots do not grow to full size and under conditions of severe boron deficiency, they remain very small and distorted and have a rough, unhealthy, greyish appearance instead of being smooth. Their surface often is wrinkled and cracked. Within the fleshy roots, hard or corky spots are found scattered throughout the roots.

Discolouration of cell walls and granulation of protoplasts of meristematic cells appear first. In the cambium, cell division continues with little differentiation into xylem and phloem and many cells are abnormally large. This leads to disturbed translocation and eventually distorted growth follows necrotic black spots. This problem is due to boron deficiency which is more common in a dry season, particularly when a long dry spell is followed by a wet period favouring rapid growth, most probably because the root system in the upper soil layer ceases to function during the dry period and thus, intake of boron is reduced.

Management

The quantity of boron needed for satisfactory management of boron deficiency varies with the nature of the soil, the soil reaction and soil moisture. Soil application of 10-15 kg of borax per hectare is recommended. Foliar application of 0.1% boric acid increases the yield, besides controlling the disorder.

❑❑❑

12

Cauliflower

a) Riceyness

It manifests in the elongation of peduncle bearing flower buds rendering curds, granular, loose and somewhat velvety. A premature initiation of floral buds is characterized by riceyness in cauliflower and is considered to be of poor quality for marketing. It is caused due to:

- Expose to the higher or lower temperature than the optimum required for a particular variety in curd development stage.
- Hereditary factors.
- Heavy dose of nitrogen and high relative humidity also contributes to riceyness.

Management

- It can be controlled by cultivation of genetically pure seeds and appropriate varieties with recommended cultural practices.
- Harvest while the curd is still firm, compact, white and fairly smooth; if it gets too mature, it will become grainy or ricey.
- Proper management of soil moisture and fertility during the development of head or curd.

b) Fuzzines

It is velvety in appearance and formed, when flower pedicles elongate. This type of curd may result due to hereditary factors or under unfavourable weather conditions or due to over maturity of curds. The disorder is both hereditary and non-hereditary. Cultivation of cauliflowers out of their normal season encourage fuzziness.

Management

- Sowing good quality seed in right season under proper cultural practices minimizes fuzziness.

c) Leafiness

This disorder is commonly observed by the formation of small thin leaves from the curds which reduce quality of curd. This disorder occurs due to prevalence of high temperature.

Management

- It can be controlled by selection of varieties according to their adaptability and seasons.

d) Buttoning

Buttoning is the development of pre-mature flower heads in cauliflower. Plants develop small curds without complete vegetative growth. It results in very small, usually loose, open, unmarketable heads. Any factor which retards vegetable growth of seedlings causes buttoning. Important causes are:

- Deficiency of nitrogen.
- Transplanting of more than 6 weeks old (over aged) seedlings.
- Low temperature.
- Root damage by insects.
- Planting of early variety late and vice versa leads to buttoning.
- Hot and dry weather which is unfavourable for vegetative growth but favourable for curd development resulting in very small sized heads.
- Inadequate moisture supply and improper plant protection measures.

Management

- Nursery should be properly looked after to avoid any check in the plant growth.

- An adequate amount of nitrogen and water should be applied.
- Do not delay transplanting. Transplanting 35-40 days old seedlings of early and mid early group of varieties.
- Cultural practices should be carried out well in time and water logging and over crowding should be avoided.

e) Blindness

It refers to the plant without terminal buds or when the growing point collapses at an early stage and the terminal buds fail to develop and plant becomes blind. It is characterized by the leaves that are large, dark green, thick and lathery owing to the accumulation of carbohydrates. Important causes are:

- Damage to the terminal growing point due to low temperature, cutworm damage or rough handling of transplants.
- It occurs in over wintered plants and any cultural practices interfering in growth of the terminal bud may lead to blindness.

Management

- Protecting young plants from low temperature exposures.
- Careful handling of the plants and good protection against insects.

f) Bracting

The bracts are underneath the pre-floral meristem, which corresponds to axillary buds. These bracts or leaves come out of the curd resulting in poor quality of curds for marketing, as they turn green or purple in colour on receiving the direct sunlight at the surface of the curd. Temperatures higher than the optimum during curding cause this disorder.

Management

- Control of temperature during curding.

g) Purple Colouring

Sometimes various pigmentations occur on the curd due to fluctuation in temperature.

h) Chlorosis

Chlorosis shows an interveinal and yellow mottling of flower leaves. The affected leaves turn bronze in colour and become stiff. In severely deficient plants, abscission of the lower leaves occur and results into small curd formation. In cauliflower, magnesium deficiency causes chlorosis, when grown on highly acidic soils.

Management

- Liming of the soil with dolomite limestone to bring the soil pH to 6.5.
- Use of fertilizers containing soluble magnesium and keeps the disorder under control.

i) Bolting

It means premature formation of seed stalks i.e. before the formation of heads or failure of leaves to form a solid head. The causes include early sowing of seeds, warm winter, extreme change in temperature, poor growth of seedlings and varietal characters.

Management

- Use resistant cultivars.
- Avoid planting/transplanting/setting plants too early into the field and prevent the stimulation of early plant growth.

j) Browning

Browning is caused due to boron deficiency. Generally, the deficiency symptoms of boron are externally visible on plants after the curd formation. In early stage, the water soaked areas appear on the stem and curd surface. As the plant grows, the stem becomes hollow with water soaked tissues covering the internal walls of the cavity. In advanced stage of deficiency, brown or pink coloured areas are seen on curd surface and therefore, it is also called brown rot or red rot or browning of the curd. Sometimes, the stem may become hollow even without brown areas on the curd. The affected curd is bitter both in fresh and cooked condition. This disorder is caused by boron deficiency.

Management

- This may be managed by application of boric acid at the rate of 15 kg/ha as soil application.
- Foliar spray of 0.25-0.5% solution of boric acid along with a sticker on 20, 35 and 50 days after transplanting.

k) Whiptail

Young plants become chlorotic and may turn white, particularly along the leaf margins and finally become cupped and wither. In extreme cases, only large bare midrib develops. This condition is commonly known as 'Whiptail'. Growing point of the plant is usually deformed and no marketable curds are produced. This disorder caused by deficiency of molybdenum causes "Whiptail syndrome" especially, in high acidic soils.

Management

- Application of lime or dolomite limestone to raise the soil pH up to 6.5.
- Apply 1.0 kg of sodium or ammonium molybdate per hectare.
- Three sprays of 0.1% ammonium molybdate solution along with a sticker on 15, 30 and 45 days after transplanting.

l) Hollow Stem

On affected plants, curling and rolling of leaves, deformed foliage and the interior pith tissue of the stem or head is collapsed or cracked and often hollow symptoms appear. The main causes are high temperatures, in combination with high levels of nitrogen and wide plant spacing also contributes greatly to the problem.

Management

- It can be corrected by close spacing and optimum use of nitrogenous fertilizers.

m) Multiple Curd

This disorder of cauliflower is marked by the appearance of a number of small button-like curds (may be 4-5 in number) in the form of a bunch. This disorder happens due to injury of terminal buds resulting in branching before curd initiation. Small curds are formed on each branch of the terminal bud.

Management

- Careful handling of the plants and following good plant protection measures so that terminal buds may not be injured.
- Exposure of plants to very low temperature should be avoided.

13

Cabbage

a) Black Petiole or Black Midrib

This is an internal disorder that has been occasionally noted in recent years. As heads approach maturity, the dorsal side of the internal leaf petioles or midribs turn black at or near the point, where the midrib is attached to the core. This disorder is believed to be associated with P-K imbalance.

Management

- Black petiole disorder can be avoided by applying balanced K and P.
- Avoiding excessive N fertilizers.
- Use of resistant cultivars.

b) Black Speck

Black leaf speck is a disorder that consists of very small to moderate size discolored lesions on the midrib and veins of the leaves. The symptoms can occur after low temperatures in the field and by harvesting over mature heads, but are usually associated with transit and storage conditions. Midribs look as if ground black pepper was sprinkled on them giving the name "Pepper spot". Black speck could be due to nutrient imbalance. Low storage temperatures followed by warmer temperatures enhance development.

Management

- High rates of potassium in the soil will reduce the severity of the disorder.
- Growing tolerant varieties.

- Storage with high CO_2 atmospheres (10%) can reduce pepper spot development on round cabbage.

c) Cabbage Splitting

Cabbage splitting is mainly a problem with early cabbage. A problem can develop, when moisture stress is followed by heavy rain. The rapid growth rate associated with rain, high temperatures and high fertility cause the splitting.

Management

Proper irrigation can prevent splitting.

d) Tip Burn

This problem can cause severe economic losses. This disorder is characterized by necrosis on leaf-margins and breakdown of plant tissues inside the head of cabbage. The problem is caused by excessive nitrogen, high temperature, water stress and calcium deficiency. Calcium can be present in the soil but its translocation to the plant is limited and hence, it may not be available to accommodate rapid growth.

Management

- Maintaining adequate moisture and uniform moisture supply.
- Maintain proper nitrogen and potassium content and soil pH.
- Use resistant or tolerant cultivars.
- Two foliar applications of CCC at 1120 or 2240 ppm combined with $CaCl_2$ @ 2g/litre or 4 applications of $CaCl_2$ alone reduced the appearance of tip burn.
- Spraying 5 times with 0.7% $CaCl_2$+50 ppm NAA.

e) Intumescence

Conspicuous, small, light yellow or gray galls up to three mm. in diameter appear chiefly on the lower side of lower leaves of cabbage growing on sandy soil. They are incited by injury from sharp particles of soil blown against the leaves during heavy winds. Following the injury, the palisade or mesophyll cells enlarge to many times their normal volume, remaining thin-walled and with relatively little cell division. The extent of the intumescence is enhanced in humid atmosphere.

f) Lightning Injury

The occurrence of lightning injury in the field is seldom noticed until some weeks after the stroke. It has then become evident as a roughly circular bare spot in which most or all of the plants are completely dead, having been killed rapidly, while at the periphery, plants show various degrees of retardation in growth. The charge is sufficiently penetrates to damage the cabbage plant without killing, it penetrates the stem at the soil line with little damage to cortex and vascular ring but permeates the more succulent pith. The cells are killed for some distance above and below the point of entry and eventually a hollow cavity results, surrounded by a dark brown lining consisting of the desiccated tissue, which has undoubtedly been broken down by enzymatic action during the process. This course of events, possibly through the release of certain hormones, stimulated adventitious root development from the inner periphery of the vascular ring and ultimately innumerable rootlets have almost completely filled the central cavity. There is a tendency, moreover, for abnormal growth of dormant buds at leaf scars above this point, while adventitious roots are formed at scars immediately below the wound.

❑❑❑

14

Broccoli

a) Floret Yellowing

Floret yellowing may be due to over maturity at harvest, high storage temperatures after harvest and / or exposure to ethylene. Any development of yellow beads ends with commercial marketability. Bead yellowing due to senescence should not be confused with the yellow to light green colour of areas of florets not exposed to light during growth, sometimes called "marginal yellowing".

Management

Proper post-harvest handling and packaging will help to minimize this problem.

b) Brown Floret

This is a disorder in which areas of florets do not develop properly, die and lead to brown discoloured areas on the broccoli head. This is thought to be caused by plant nutritional imbalances but also may be due to feeding damage on florets from insects.

c) Internal Tip Burn

Tip burn causes leaf margins to turn brown and leaves to be buried in the head. This has been ascribed to poor water movement with the plant.

Management

Maintaining adequate moisture level and uniform moisture supply.

d) Broccoli Heat Injury

High temperature causes the heads of broccoli to be rough, with uneven head size.

Management

Use heat resistant or tolerant cultivars.

❑❑❑

15

Lettuce

a) Tip Burn

This is a physiological disorder caused by the plant's inability to obtain sufficient calcium due to rapid plant growth, excessive fertilization rates, or uneven water availability. Avoid excess fertilization. Irrigate with drip irrigation systems.

b) Russet Spotting

A post- harvest disorder characterized by localised, spot like lesions that may start either in the epidermis or in the mesophyll which, in advance stage, may show discolouration of vascular tissue and collapse of mesophyll cells resulting in pit-like depression. Ethylene accumulation to the tune of 0.1 ppm concentration is sufficient to cause this disorder during a normal transit period of 5-8 days at 5°C.

c) Classiness

Classiness of lettuce is characterized by water soaked spots, mostly along the leaf margins and bounded by the veins. Affected tissue may die and appearance is marred. This disorder tends to occur under low light intensities. The plants take up more water than they are able to loss by transpiration.

Management

Control measures include raising the temperature at night, compacting the soil and increasing the supply of nitrogen and mineral salts. Cultivars differ widely in their susceptibility to glassiness.

d) Heart -Leaf Injury

Heart-Leaf injury of lettuce, in which the innermost leaves turn brown, is usually the result of an increased concentration of carbon dioxide or due to lack of oxygen. Avoidance of this storage disorder depends on good temperature control to reduce the rate of respiration and appropriate air exchange rate to prevent accumulation of respiratory gases.

e) Brown Stain

It appears as large yellow reddish brown depressed spots on the mid ribs and reddish brown streaks on the other parts also. Brown stain is caused by exposure to 3% CO_2 at low temperature.

❏❏❏

16

Celery

a) Cracked Stem

Cracked stem, sometimes called brown check is related to a deficiency of boron. There is brownish mottling along the leaf margins, followed by the brittleness of the petiole. The affected tissues collapse, become light yellow and later form a corky layer. This disorder begins as a crack of the inner petiole surface below the first leaflet and extends to the stalk. The exposed tissue becomes brown and susceptible to disease infection. Crosswise cracks appear in the outer layers of the petiole and tissues surrounding the cracks turn brown. The roots also turn brown and die.

Management:

The disorder can be corrected by spraying of 625 g of boric acid in 1200 litres of water per hectare or by the application of borax @ 12 kg per hectare in the soil at the base of the plant.

b) Black Heart

It occurs in most regions, where celery is grown. It is the young leaves that show tip burn symptoms first, then spreads quickly to most of the heart tissues. This is followed by drying, blackening and in severe cases, the killing of the entire heart. It is caused by the calcium deficiency in the young leaves, which causes the black heart i.e. inner leaves and the growing point to turn black and the affected tissues are invaded by the organisms viz. bacterial soft rot, resulting in black, water soaked centers. The malady is not confined to the inner leaves but may also affect the outer leaves which may turn yellow. The disorder is very similar to tip burn in lettuce and cabbage, appearing when weather conditions favour a sudden growth surge and available calcium cannot meet plant requirements. This disorder is reported to be accentuated by high temperature, rapid growth and water stress. It is a growth related disorder, as plants with high nitrogen rates are more susceptible.

Management

- Spraying of Calcium nitrate @ 6.5 to 18.75 kg in 1200 litres of water per hectare to the growing point.

c) Blanching

It is the exclusion of light from the leaf stalks of the growing plants and make them devoid of chlorophyll. This process is called blanching. Proper blanching makes the crop crisp, reduce acid flavour, improves flavour and keeps the plants tender with elongated stalk which adds to the quality and forms a good salad. It is done by wrapping with dark coloured brown paper strip, old newspaper and with board around the leaf stalk. By earthing up in gradual stages for three times as the plants grow, covering the basal portion, leaving the leaves exposed. It is also done when the plants attain 30 cm high, the outer older leaves are removed and the remaining leaves are tied together. Blanching is done in moist warm weather, soil-blanching helps in protecting the crop from frost injury in colder place only.

d) Petiole Pithiness

Petiole pithiness is a widespread disorder in stalk celery. Anatomically, this disorder is characterized by a breakdown of the parenchyma cells, leaving large air spaces in the parenchyma which under severe conditions may merge with one another and form a hollow petiole. Several types of root stress have been found to cause pithiness: water deprivation acted rapidly; whereas flooding and nutrient deficiency exerted a continuing effect.

Management

Petiole pithiness is completely eliminated, when fertilization with low nitrogen concentration during the first two-third of the growth period was followed by nitrogen-rich fertilization until harvest.

e) Chlorosis

This is characterized by yellowing of older leaves due to magnesium deficiency. It occurs in cultivars that are inefficient with respect to magnesium uptake and metabolism. It is due to a single recessive gene, but high calcium in the plant will intensify the disorder.

Management

This disorder can be corrected by spraying magnesium sulphate @ 2.5 kg in 1200 litres of water per hectare on the foliage twice with 7 days intervals.

❑❑❑

17

Peas

a) Hollow Heart

Hollow heart is a disorder characterized by a sunken area in the adaxial surface of the cotyledons of the pea seed. This condition is thought to be brought about by rapid drying as the seed approaches maturity. The disorder causes poor seedling vigour and poor stands as the cotyledons are often attacked by soil borne damping-off pathogens.

b) 'Blonde' Peas

'Blonde' pea is a disorder that is characterized by variations of light green and yellow (blonde) peas in green peas harvested for freezing or canning. The condition is brought about by shading of the pea pods by excessive vine and leaf growth. Also, the condition may be caused by lack of sunshine and cloudy conditions during pod filling.

c) 'Bleached' Peas

'Bleached' pea is a disorder, sometimes confused with 'Blonde', that is caused by excess moisture after the peas have reached dry seed maturity. Apparently, the dry peas take up moisture leading to enzyme activation and a breakdown of chlorophyll. The condition causes colour variations and is a market downgrading factor in dry peas. Some cultivars have a degree of resistance but when conditions are favourable for bleaching, all cultivars can be affected.

d) Cold Injury

Cold injury can occur, when temperatures drop below 0°C, particularly if pea plants have been growing at high temperatures. Cold-hardened plants are less susceptible to injury. Plants growing in low areas in the field and in areas of poor air movement are more prone to injury. Symptoms include killing of the growing tips on very young growth; roughened, jagged edges or leathery, bilobed young leaves; and long, water-soaked, translucent lesions between the main veins. These lesions may turn necrotic on the underside of older leaves. If frost occurs at the green-harvest stage, lacy white lesions form on pods, resulting in uneven quality. Tall cultivars are more susceptible to frost injury than shorter ones.

e) Herbicide Injury

Herbicide injury may occur through improper use of herbicide on a pea crop itself as a result of herbicide drift from nearby crops, or from herbicide residues in the soil resulting from application to a previous crop. Symptoms vary with the herbicide and may include yellowing and stunting of the plant, mild to severe distortion and epinasty of the stems and leaves, abortion of flowers, pod abscission, reduced seed set, foliage chlorosis and necrosis of the lower leaves. Proper use of herbicides reduces this problem, but special precautions may be needed to prevent drift between fields.

f) Water Congestion

Water congestion may result from adverse environmental conditions, such as high humidity, high temperature and high soil moisture, particularly on clay soil. If the conditions that foster this disorder persist, considerable yield loss may occur, because of the reduction in vigour that accompanies the loss of photosynthetic surfaces. Small, water-soaked spots first appear near the outer edge and on the underside of pea foliage. The terminal portions of leaves and stipules become completely water-soaked as the spots increase in size. The affected tissue first appears darker green. But as gradual death occurs, which progresses from the outer most tip and edge of the foliage, the necrotic tissue takes on a shrunken, dry, straw-coloured appearance. Most of the foliage at one to several nodes may be affected, reducing vigour and yield.

Management

Improved weather conditions generally arrest the development of this disorder.

❑❑❑

18

Cucurbits

a) Blossom end Rot

It is a physiological disorder observed in many cucurbits as well as other crops. Blossom end rot is usually the result of inadequate or uneven irrigation, high humidity, or other factors that slow the movement of water through the plant. Since calcium is taken into the plant with the transpiration stream (water), slow water movement can often lead to temporary calcium deficiencies, resulting in blossom end rot.

Management

Provide adequate calcium and proper irrigation. Do not use high levels of ammonia fertilizer, which can aggravate this problem.

b) Drought Stress

Cucurbits are particularly sensitive to drought. Fruits are typically 85% to 90% of water and can suffer under drought conditions. Pumpkins often produce long vines with many leaves and can transpire large quantities of water during hot summer days. Severe drought stress affects fruit development, resulting in unmarketable produce. Affected cucumber fruits may appear curled, distorted, or tapered at the blossom end; pumpkins become soft and wrinkled. In addition, drought-stressed pumpkins fail to gain appropriate size, which affects yield. A loss of foliage during drought will also result in sunburn of the fruit.

Management

Irrigate, whenever necessary.

c) Flood Damage

Flood damage symptoms often appear as nutrient deficiencies or as a generalized yellowing. Prolonged exposure to flooded soils will result in anaerobic (low oxygen) conditions for plant roots, eventually causing death. When large numbers of roots die, the plant is often unable to take up sufficient nutrients, resulting in nutrient deficiencies.

Management

While damage from flooding is often unavoidable, planting in raised beds will improve drainage.

d) Preponderance of Staminate (Male) Flowers and Low Fruit Set

In monoecious sex form, staminate (male) and pistillate (female) flowers are separately produced in the same plant, as in cucumber, bottle gourd, pumpkin, summer squash, watermelon, ash gourd, bitter gourd, ridge gourd, sponge gourd, snake gourd and long melon. In andromonoecious sex form, staminate and perfect flowers (hermaphrodite) are produced separately in the same plant, as in muskmelon and some cucumber cultivars. In both these sex forms, yield depends on the number of pistillate/perfect flowers. Normally, male flowers appear in the first 4-6 flowering nodes and later pistillate/perfect flowers appear in few nodes. In some cases, pistillate flowers appear early as seen in pumpkin but it does not indicate that number of pistillate flowers in that plant would be of greater proportion than the staminate flowers. In both monoecious and andromonoecious sex forms, staminate flowers are produced in far greater flowers in the plant should be narrowed down to increase the number of fruits and eventually yield per plant.

Management

- Rampant vegetative growth should be avoided, as it promotes maleness.
- Excessive nitrogen application and over irrigation should be avoided, as it generally promotes greater number of staminate flowers.
- Pruning of vines: In muskmelon, pruning is quite successful in increasing the yield. Generally, secondary branches are pinched

off up to 6-7 nodes on emergence and tertiary branches are allowed to grow. Perfect/pistillate flowers are allowed to be borne on these branches to set fruits. These tertiary branches are detopped at 2 leaves above the first perfect/pistillate flowers.

In watermelon, the main shoot is pinched beyond 5th or 6th node and 3-4 well- developed primary branches are allowed which bear the pistillate flowers.

In pumpkin and bottle gourd, cutting of extravagant vine growth helps in narrowing the ratio of staminate: pistillate flowers.

- Exogenous application of growth substances and other chemicals: Application of growth substances and some nutrient elements at 2 to 4 true-leaf stage can alter the sequences of flowering and narrows the ratio of staminate: pistillate/perfect flowers in desired direction. Application of chemicals may be done twice or thrice at an interval of 7 days. Effective concentration of the growth substances and nutrient elements for different cucurbits are :

  Muskmelon: Maleic hydrazide (MH) @ 200 ppm.

  Watermelon: 2,3,5 tri-iodobenzoic acid (TIBA) @ 50 ppm, MH @ 50 ppm and Boron @ 2 ppm.

  Pumpkin: 2 chloroethylphosphonic acid (ethrel or ethephon) @ 250 ppm.

  Summer squash: ethrel @ 600 ppm.

  Ridge gourd: ethrel @ 300 ppm

  Bottlegourd: MH @ 50 ppm, Boron @ 3 ppm, Calcium @ 5 ppm, 2-chlorophenoxyacetic acid (2-CPA) @ 100-200 ppm.

  Snake gourd: ethrel @ 100 ppm.

  Bitter gourd: Naphthalene acetic acid (NAA) @ 100 ppm, 6-benzylamino purine (BA) 25 @ ppm, ethrel @ 200 ppm, 2-CPA @ 100-200 ppm.

  Sponge gourd: ethrel @ 250 ppm.

  Cucumber: ethrel @ 250 ppm, MH @ 100 ppm.

e) Hollow Heart

It is the formation of a hollow cavity inside some cucurbit fruits. This disorder can result from a number of factors, including low boron levels, genetics and uneven water management. Although not outwardly visible, hollow heart makes fruit unmarketable.

Management

Avoid varieties with a tendency to exhibit hollow heart. Ensure that boron levels in the soil are adequate; however, be careful not to over fertilize. Follow recommended plant spacing and avoid erratic irrigation.

f) Pillow

It is a fruit disorder of processing cucumber due to low calcium level in the tissue. In this disorder, an abnormal white like porous textured tissue is formed in the mesocarp of the fleshy harvested fruits. Vascular tissue with some pillow areas may collapse and become necrotic.

g) Leaf Silvering

It is a physiological disorder of summer squash. This disorder occurs due to moisture scarcity. The leaves become silver coloured and contain less chlorophyll; photosynthesis is hampered in the silvered leaves.

h) Unfruitfulness in Pointed Gourd

Pointed gourd is a dioceous cucurbit. So, male and female plants are separate. Female plants to ensure adequate pollination, fertilization and fruit set. A common problem is met with where pistillate flowers in female plants are shed due to lack of pollination and fertilization.

Management

Male plants must be grown in the field along with the female plants at the rates of 10-12 male plants per 100 female plants to ensure adequate pollination and fruit set. Hand pollination may be done successfully to achieve fruit set. Hand pollination to the female flowers should be done in the early morning hours because stigma receptivity decreases with an advancement of the day.

i) Delay in Fruit Ripening

This problem is particularly important in muskmelon and watermelon. Delay in ripening is sometimes associated with less sweetness and cracking of fruits which occur due to high moisture level and temperature fluctuation at ripening stage.

Management

Irrigation should be stopped at the ripening stage to hasten ripening. Sowing time should be adjusted in such a way that fruits ripe in hot and rainless condition which hastens ripening and at the same time improves sweetness of the fruits.

j) Light Belly Colour

This disorder is characterized by the under surface of cucumber fruit remaining light in colour instead of turning dark green. Commonly occurs on fruit lying on cool, moist soil.

Management

Avoid excessive nitrogen. Can be managed by avoiding luxuriant vine growth.

k) Measles

Symptoms are most evident on smooth-skinned melons and cucumbers. Small brown spots are scattered over the surface of the fruit. The spots are superficial and do not penetrate beyond the outer epidermal layers of the fruit. These spots also may occur on leaves and stems. The guttation droplets develop high concentrations of salts which burn the epidermis. Measles spots occur, where a guttation droplet had formed.

l) Rind Necrosis

It is an internal disorder of the watermelon rind. Symptoms are brown, corky, or mealy textured spots on the rind which may enlarge to form large bands of discolouration that rarely extend into the flesh. The cause of rind necrosis is unknown. Bacterial infection has been reported to be a cause, although similar bacteria are found in healthy melons. Drought stress also is reported to predispose melons to rind necrosis.

m) Leaf Aging

Discoloration, bronzing, whitening, brittleness and leaf tearing are common on older leaves of many vegetables, particularly cucurbits. In cucurbits, the oldest leaves (crown leaves) in watermelons, squash, cucumbers and pumpkins will have large areas that are discoloured (white, tan, or bronze). These leaves will be brittle to touch and may start to tear or shred with high winds and storms. This condition is common in cucurbit crops but also can be found in other vegetables such as tomatoes and can be due to a number of leaf aging factors including mineral nutrient scavenging (export of mobile nutrients from oldest leaves to newer leaves), ozone air pollution damage, chemical phytotoxicity (copper fungicides for example), repeated stress cycles and wind injury. Leaf cells that die will leak their contents, releasing enzymes and oxidizing chemicals affecting nearby cells, thus accelerating the ageing process. This results in large patches of dead leaf cells that will dry, making the leaf feel brittle. If leaf veins are damaged, water and food transport will be compromised, accelerating leaf decline. This leaf aging is not to be confused with damage from mite feeding which is also concentrated on oldest leaves.

n) Oedema

Oedema is a physiological disorder of cucurbits and is most frequently found with pumpkins and winter squash that are subjected to undo moisture stress. This is most often associated with uneven availability of moisture, when immature fruit are enlarging. On pumpkin, rinds the severity may be enough to make the fruit unmarketable. The fruit surface is often raised with circular-shaped lesions that are corky or 'crusty' in appearance and may appear irregularly over the entire surface, or be limited to, for example, the shoulder side exposed to direct sunlight. The crusty appearance is similar to the appearance of scab on hard-shelled fruit, except that the oedema lesions never appear crater-like or shrunken. Excessive loss of foliage due to severe powdery mildew can cause plant stress and oedema-like linear cracks can also develop on the surface.

❑❑❑

19

French Bean

a) Delay in Flower Initiation and Development:

At temperature below 10°C, fertilization may not occur and produce small and misshapened pods. At high temperature (above 35°C), blossoms drop and ovules abort.

Management

Planting crop at suitable time and place are remedial measures.

b) Transverse Cotyledon Cracking

It takes place, when dry seeds are sown in wet soils. Seed moisture content below 12% increases cracking of cotyledon.

Management

Use of resistant cultivars with hard seed coat, optimum seed moisture content and planting of crop at appropriate time are effective measures.

c) Hypocotyl Necrosis

It is due to low Ca content in seed after germination. Soils rich in Ca and Mg overcome the problem.

❑❑❑

IV. Spices and Plantation Crops

1

Coconut

a) Button Shedding

Abnormal button shedding and immature nut fall causes enormous reduction in the nut yield of coconut. Falling of undeveloped, fertilized female flowers is called button shedding in coconut. Button shedding during the first to four months of development followed by shedding of immature nuts is a common phenomenon in coconut. Normally 50-60 per cent buttons are shed in coconuts which will extent up to 70 per cent during summer. The remaining buttons develop into nuts. Button shedding of more than 60-70 per cent is called excessive/abnormal button shedding which will affect the nut yield of coconut palms. Button shedding and immature nut fall is caused due to several factors viz., genetic causes, soil and climatic factors, nutritional deficiencies, hormonal deficiency and infestation of insects and pathogens.

Factors of affecting button shedding

(i) Genetic causes

Genetic characters of mother palm is one of the prime reasons for button shedding in coconut. Even in well maintained coconut gardens, there will be a heavy button shedding and immature nut fall. This is due to the poor nut yielding capacity.

(ii) Selection of seedlings

Selection of good yielding mother palms is very important for collection of seed nuts for raising coconut seedlings.

(iii) Sufficient drainage

Providing sufficient drainage facilities to coconut trees will be helpful to remove the harmful chemical substances from the soil.

(iv) Soil pH

Coconut is well grown in soils having pH of 5.5-8.0. When the soil acidity or alkalinity is high, there will be more button shedding and immature nut fall in coconut.

Management

Application of 2 kg lime/tree will be helpful to rectify the excessive acidity in soil. For alkaline soils, application of 4 kg gypsum/tree is recommended to reduce soil pH and for correcting the excessive soil alkalinity. Applications of green leaf manure and farm yard manures are also useful. Growing of coconut in rocky substrata or gravelly soils will cause heavy button shedding. Planting of coconut in such soils should be avoided.

(v) Drought or Excessive Irrigation

Drought condition or water shortage for a longer period also causes heavy button shedding in coconut. Coconut tree requires about 60-80 litres of water per day. Irrigation during summer months is highly essential to reduce the button shedding in coconut. Sudden alteration in soil temperature and soil moisture also leads to heavy button shedding and immature nut fall in coconut. Not only the drought condition, but also the water stagnation is harmful to coconut. If there is no sufficient drainage facilities, aeration to coconut roots will be very much affected. Chemical changes in sulphides and water soluble iron and aluminium complex substances in poorly drained soils affect the nutrient uptake in coconut.

Management

Between two rows of coconut trees, during rainy period, burying of coconut husks in 6m x 1.5m x 45 cm trenches in 2 layers and covering with soil will be useful to conserve moisture. Application of coir pith and dried spathes in coconut basins is also useful. Hence, irrigation management/soil moisture conservation is very important to reduce the button shedding in coconut.

(vi) Nutritional Deficiency

An adult coconut tree utilizes 540 g nitrogen, 260 g phosphorus and 820 g potassium from the soil in a year. To compensate this nutritional loss, adequate nutrient applications are to be done.

Management

Coconut trees are to be applied at the rate of 1.3 kg urea, 2 kg super phosphate and 2 kg potash along with 50 kg FYM/palm/year. Application of additional dose of 1 kg potash (3 kg potash) reduces barren nuts and button shedding. Blanket application of 1 kg coconut micronutrient mixture per palm per year is also suggested. Application of 200g zinc sulphate/palm/year is recommended for coconut trees with heavy button shedding.

(vii) Poor pollination

Poor pollination is also one of the factors responsible for button shedding in coconut. It causes lesser rate of setting of buttons into nuts.

Management

Honey bees improve pollination in coconut gardens. Keeping 10-15 bee hives in one hectare of coconut garden will be useful to improve the cross pollination. Bee keeping in coconut gardens also gives additional income to the farmers.

(viii) Plant Growth Regulators

Plant growth regulators are highly essential for the development of fertilized female flowers into nuts. In some coconut trees, the production of plant growth regulators is less.

Management

Spraying of NAA (Planofix) @ 1ml in 2 litres of water for 30 days after the opening of inflorescence in coconut trees will reduce the button shedding. This NAA spray is found to increase 10 per cent nut setting in coconut.

(ix) Insects Infestation

Infestation of sucking pests like scales, mealy bugs and eriophyid mite on coconut bunches also causes heavy button shedding in coconut. Recent outbreak of the eriophyid mite has caused severe loss to the coconut farmers by means of heavy button shedding/immature nut fall and poor nut development.

Management

Root feeding of 30 ml TNAU - agrobiocide followed by 15 ml Carbosulfan + 15 ml water at 45 days interval and repeated with TNAU - agrobiocide after another 45 days controls the eriophyid mite. Spraying of 1.5 ml monocrotophos or 4 ml methyl dematon in 1 litre of water along with adhesive material is effective against sucking pests.

(x) Plant Pathogens

Microbes/plant pathogens are also the major causes for the button shedding and immature nut fall in coconut. Button shedding due to pathogen infection is normally observed during rainy season. Fungi like *Phytophthora* sp, *Colletotricum* sp, *Botryodiplodia theobromae, Botryosphaeria rhodina Pestalosia* sp, are reported to be associated with button shedding and immature nut fall in coconut. One to four months old buttons are more susceptible to pathogen infection.

Management

Spraying of mancozeb at 0.2% is effective in reducing the button shedding. Though various factors are responsible for button shedding in coconut, the exact cause is to be identified and appropriate remedial measures have to be taken upto increase the coconut yield.

b) Barren Nuts in Coconut

In general, the nut malformations are more in the case of hybrids than in the tall varieties. On an average, in any coconut garden, there will be three per cent barren nut production and if this percentage is on increase, the control measures are ought to be taken. The production of barren nuts is related to genetic as well as nutritional. The seedlings that does not conform to the quality standards may produce a continuous barren nut formations and cannot be rectified at all. Besides, the seed nuts that are collected from mother palms of such characters may even show up the disorder.

Management

- Nut malformations are often related to the deficiency of the potash and boron nutrition. In many cases, the application of potash @ 2 kg per palm per year over and above the normal recommendation has shown good improvement. Similarly, application of borax at the rate of 100-250 per palm per year depending upon the severity of the problem produced desirable results. In addition to the above, the organic matter build up of the soil by means of liberal dosage of FYM or pig or poultry manure will improve the yielding ability of palms and reduce the nut malformations. Apply extra one kg of K_2O with 200 g of Borax/palm over and above the usual dosage of fertilizer to correct the barren nuts in coconut.

- Regularly survey for pest and diseases
- Select seedlings of 9 to 12 months after planting. Seedlings, which have germinated earlier, having good girth at collar and early splitting of leaflets, should be selected for planting.

c) Crown Choke

Emergence of shorter leaves with deformed and crinkled leaflets is the first symptom of the disease. Leaflets often show severe tip necrosis. Number of leaflets reduce gradually and hooking is also seen. Affected leaflets do not unfurl properly and in many cases give a choked appearance to the frond. As the disease progresses stick-like leafstalks emerge. The inner leaves crowd round the bud and prevent normal unfurling of the flag leaf. Necrosis may proceed to the deeper tissues and kill the meristems. Premature nut fall is seen and gradual reduction in yield is noticed. Inflorescence emergence is hindered resulting in yield loss.

Management:

- Soil application of borax @ 50g / palm at half yearly intervals, during Feb-March and Aug-Sept., improves the condition of the palm.
- Making slits in the tight extension of the petioles and opening out the crown would help in the normal emergence of fronds.

d) Crown Cleaning

Each leaf will have a central stalk of 6m in length. The base of the stalk is surrounded by a fibrous network of light brown stipules. In this situation, crown should be cleaned, pruned and stripped for effective sanitation, aeration and sun light to filter. Crown cleaning is to be done once in a year. During the crown cleaning, the dried fronds, stalks, spath, spadices are removed. By frequently cleaning the crown, more sun light enters among the palms, is effective, in air circulation and helps in pollination and the insect pest damage is reduced since the crown cleaning restricts the hiding place. The fibrous network materials can be used as the mulch around the tree to check the evapotranspiration loss.

e) Pencil Point Disorder

Tapering wilt or pencil point disorder of coconut is common in the gardens. The pencil point disorder occurs on an average of 5-10 per cent in the rainfed gardens and limits the productivity to a great extent. This disorder, as its name implies, is characterized by the pencil tip at the apex of the crown. The trunk starts tapering towards the crown with simultaneous reduction in the number and size of leaves and the appearance of the palm becomes unhealthy and debilitates with age.

There are two types of pencil point disorders. In the first one, the palm will be all along yielding very well and suddenly the trunk starts tapering. The size of the photosynthetic surface as well as the total number of functional leaves, production of spathes and female buttons get reduced and if left uncared, the palm will become unproductive. Such palms are affected by micronutrient deficiencies. The second type of the disorder develops at the early stage itself indicating the genetically poor materials and difficult to be recovered. Care may be taken to select quality seedlings for fresh planting.

Management

To overcome the disorder, it is recommended to apply 225 g in each of sulphates of zinc, copper, iron, manganese and borax and 10g of Ammonium Molybdate. These nutrients may be dissolved in 40 litres of water and drenched in the six feet radius palm basins formed around the tree. Quarterly applications seem to benefit the palm for rejuvenation. The micronutrients may be applied at the beginning of the symptoms for recovery.

f) Drought Injury

The common events due to drought are yellowing, drying and drooping of leaves in the outer whorl of the crown. Depending upon the severity of drought, these symptoms are extended to the inner leaves also. Literally, the palm undergoes dehydration and loses capacity to form normal female flower production; as a result, less female flowers are produced; button shrinking, button and nut shedding is observed. Abortion of inflorescence and buckling of bunches also occur. If the drought is acute, wilting and death of palms is common especially in unirrigated plantations. If the planting is shallow, the drought effect is rather fast. The drought may also predispose severe incidence of insects (for example, scales, mealy bugs) and disease (for example basal stem rot, gray leaf spot, leaf blight) aggravating the problem.

Management

Management of drought is important in view of the perennial nature of the crop. Measures of reducing the transpiration loss of water (by removal of 3-5 older leaves), moisture conservation in basins (husk burial), mulching, cover cropping, increasing water holding capacity of soil (organic manure) and control of insects and diseases are suggested to tide over the drought conditions.

g) Flooding

Prolonged flooding of the young palms can render yellowing of leaves and make the plants look sickly. Excess flooding during monsoon followed by severe dry condition may also lead to a shock in young palms resulting in their death.

Management

Adequate drainage provision in coconut gardens is essential to manage the ill effect of flooding.

h) Foliar Yellowing

Magnesium deficiency leads to loss of chlorophyll in leaves. Hence, the marked visual symptom is the yellow discolouration of the leaflets of the oldest frond and the yellowing spread from the leaflet tips towards the rachis of the leaf. The yellowing becomes intense on part of the palms exposed to the sun. The discolouration is less pronounced along the mid-vein of leaflets. With acute deficiency of the mineral, the leaflets tend to lose almost all its pigments. Necrosis is seen on the tip of leaflets which turn reddish brown and translucent patches appear on the yellow leaflets. Magnesium deficiency affects the number of nuts per tree.

Magnesium deficiency can occur in soils with poor exchangeable magnesium. In coconut, it seems that the deficiency can also occur due to high potassium level. An antagonism between Mg and K exists and excess K can induce Mg deficiency. Such induced Mg deficiency may be more common in Dwarf and DXT hybrid coconuts than in Talls. With the correction of K level, there would be a correlation between Mg levels and the number of green leaves. Soil groups of Kerala are said to be deficient in Mg and the sandy loam soils of coastal alluvium were more deficient.

Management

Magnesium fertilizer, usually magnesium sulphate @ 250 g per plant, can effectively correct the deficiency disorder.

i) Rubbery Copra

Rubbery copra disorder of coconut is reported to be due to acute deficiency of sulphur. Orange yellow colour of young as well as old leaves, weakening of the rachis followed by leaflet necrosis are the major sulphur deficiency symptoms. With the increase in the age of the leaves, chlorosis and necrosis increases and even the first leaf may be yellowed. The number of leaves and nuts are reduced. The nuts are small with apparently normal kernel which upon drying become soft, flexible, leathery and brown (referred to as rubbery copra) with poor quality. In rubber copra, oil content is less with high nitrogen, ash and sugar contents and the oil has high iodine value with low saponification value as compared to the oil from normal copra.

Management

Sulphur deficiency can be corrected by the application of sulphur or sulphur bearing fertilizers. Hence, inclusion of any one of the sulphur bearing fertilizers in the fertilizer schedule is necessary; for example, regular addition of magnesium sulphate (13 per cent S) or Ammophos (15 per cent S) could raise the foliar S to a desired level on a S-deficient sandy soil.

j) Lightning Injury

Coconuts are particularly susceptible due to their great height, but other tall palms are equally susceptible. The most important effect is the extreme heat, speed and shock waves generated by the bolt. The electrical current produces temperatures in excess of 50,000°F in a millionth of a second. The velocity is about 1/10 the speed of light making the trip from cloud to ground, through or around the plant and back to the cloud in about 100 millionths of a second. A direct lightening strike on a palm is usually fatal. The symptoms of lightening as follows:

- Symptom onset occurs within days of recent thunderstorm.
- A sudden collapse of the crown occurs, usually beginning with the older, lower fronds.
- The collapsed fronds are at first green but become brown and necrotic within days or in two to three weeks.
- The crown palms is sometimes broken and collapsed at its base.
- An upper hollowed trunk may at first be filled with slush from the overheated shattered tissue.
- Odourless wound during the first few days of the injury and perhaps weeks thereafter.
- Longitudinal gash on the trunk and/or crownshaft.
- Short holes in the trunk.
- Bleeding of the trunk.
- Fallen nuts.
- Injury to understory plants.
- Several palms can be killed by a single strike if they are in close proximity. The lightning can reach from one palm to the other in a grouping. If this occurs, it is possible that palms directly hit will die very quickly; whereas the other palms are more likely to linger for months before dying.

Management

Lightning injuries are usually fatal to palms although some may survive six months or longer following a strike. The injury cannot be treated and prevention by mounting lightning grounds within palm canopies is impractical for all but the most valuable palms.

The lightning affected palms attract insects like rhinoceros beetle and red palm weevil. The partially affected palms show some salmon pink exudation from the trunk and certain insects like *Xyleborus* sp. can invade. Obviously, there is no strong measure to manage the lightning injury. However, sanitation of dead palms should be strictly followed by prompt removal of such palms to avoid pest infestation and fungal colonization.

❑❑❑

2

Arecanut

a) Band

This is an abnormal condition of the palm rather than a disease and no biotic agents have so far been found to be associated with this disorder, known as 'band' in Marati, which means barren because the palms cease to produce fruits and 'hidimundige' in Kannada (a sort of constriction at the crown region), this disorder is prevalent in many areca-growing tracts with different growing conditions, especially the type of soil. The disorder was noticed even before 1889 in parts of South Maharashtra.

The main symptom of affected palm is the reduction is size of the leaves having characteristic dark green colour. The leaflets are brittle, leathery and crinkled with wavy margin. The leaves remain closely attached to the stem and fail to open the natural way, which results in the formation of 'rosette' shape of the crown. Sometimes, multiple shoots are developed or the newly formed shoots may emerge through the sides of tightly folded lower leaves. Gradual reduction in intermodal length and tapering of stem are other crown symptoms. Roots become brittle and crinkled and are poorly developed. In the acute stages, the production of inflorescence is arrested and the palms remain unproductive indefinitely or recover itself without any treatment. The inflorescence if produced will be small and malformed. Biotic agents are not associated with this disorder which is not contagious and is also not transmitted through any other means. The nutrient status of soils of healthy and diseased gardens did not show any significant difference. The disease is prevalent in gardens having poor drainage and soil aeration and low soil fertility.

Management

The palms with band symptoms will not respond to chemical method of control since no biotic agents are associated with the disease. Incidence of band can be reduced by proper cultural operations and improvement of drainage. To improve the soil aeration, hard soil strata, if present, has to be removed. This in turn will improve the formation of better root system. Incorporation of a mixture of equal quantities of copper sulphate and lime to the soil and application of borax @ 25 g/ palm will improve the condition of the affected palms.

b) Sun Scorch and Stem Breaking

The stem of border palms, which are exposed to solar radiation are affected by this disorder which was reported decades ago from Dakshina Kannada district of Karnataka. The severely scorched palms breakaway due to heavy wind during southwest monsoon season resulting in complete loss. First visual symptom is the development of golden yellow spots on the exposed side of the stem. Gradually, the affected portion becomes dark brown in colour followed by development of 1 -3 cm deep longitudinal fissures. Further, decline is by the growth of saprophytic organisms and infestation by small insects. At this stage, the stem breaks due to heavy wind during monsoon season.

Management

The palms can be protected from scorching and subsequent breaking by trailing pepper vines or raising shade trees on the south-western side of the garden or tying the trunk with palm leaves. Aligning the rows of palms in north-south direction and planting on quincunx system also lowers the incidence of sun scorch.

c) Nut Splitting

Areca palms in the age group of 10 – 25 years are more prone to nut splitting. Biotic agents have not been found to be associated with this disorder. The characteristic symptom is the premature yellowing of nuts followed by splitting which may occur from the perianth or distal end or from both ends of the fruits. Sometimes, the kernel also gets split and is exposed. This disorder becomes serious during the rainy season, as sudden flush of water after a long period of drought is attributed as the cause of the disease. Sometimes, the water stress in soil also disturbs the normal development of pericarp and kernel. Potash deficiency has also been reported as a probable cause of this abnormality.

Management

Application of potash fertilizers and spraying of boric acid @ 2 g/ litre of water before the expression of symptoms is found to be effective treatments and are still being practiced. In gardens having high water table, the drainage has to be improved.

d) Crown Choking

Crown choking can be identified at initial stages with appearance of dark green colour of leaves and reduction in size of old leaves.

Management

Soil application of zinc sulphate @ 10g per palm is ideal if disorder symptoms are in initial stages. Spraying of 0.5% zinc sulphate (5g per litre of water) mainly on fresh foliage can be done so that new leaves will emerge normally.

e) Yellow Leaf Syndrome

Yellowing of leaves in arecanut palms has been recorded as early as 1940. The Departments of Agriculture and Horticulture in Karnataka and later, the Central Plantation Crops Research Institute, Kasargodu, made some attempts to find out the cause of the problem and the extent of its spread, by way of survey and control measures through plant protection and fertilizer application.

Typical symptoms are characterized by the start of yellowing at tips and spreading to the middle, along distinct boundaries on the leaflets. The entire leaf turns yellow with a long green patch in the centre of the leaf. New leaves become shorter, closely bunched and sometimes wilt and shred. There is tapering of stem with reduction of girth at the top. The distance between nodes and the size of emerging leaves are highly reduced leading to stag head. The crown/head breaks off and falls. Roots become discoloured, suberised and absorption of nutrients is affected. They often decay and die. New roots are not produced. Simultaneously, the nuts in these plants become small, shrivelled and kernel discoloured and fall off. Symptoms start during the beginning of monsoon and reach peak during September – October and later subside due to fall of affected leaves. Yield in the affected palms is reduced by 30-40 per cent per year and in 3-4 years, the plants become unproductive or the crown falls off. The leaves many times turn yellow, when sufficient nitrogen is not given or mite infestation or root grub or other pest attack is present. Here, a general diffuse yellowing, is confined to the leaves attacked by the pests. Nitrogen starved plants also show diffused yellow colour in all the leaves. The yellowing symptoms do not show distinct margins, in all these cases.

The cause for manifestation of the yellow leaf symptoms has been reasoned out to be due to various factors. Some had thought it to be a disease caused by virus or mycoplasma. Besides, involvement of bacteria, fungi, nematodes, root grubs and mites are suspected. Some scientists held the view that the problem is more due to want of proper management of garden and nutritional imbalances caused by excess moisture and inadequate drainage, leading to toxicity of aluminium, iron and manganese found in greater quantity in malnad soils (and deficiency of phosphorus, magnesium and zinc). Since no definite causative factors or pathogen or pest are identified till recently, it is decided to term the yellow leaf problem as Yellow Leaf Syndrome.

❑❑❑

3

Rubber

a) Tapping Panel Dryness (TPD)

It is the only major physiological disorder affecting the rubber tree. This syndrome is characterized by partial or complete drying up of the panel (i.e. no production of latex) after a period of prolonged and late dripping of latex for a few days/weeks. Some trees become dry even without late dripping. In some dry trees, tumors and necrotic bark tissues develop along the bark. The affected soft bark becomes light brown in colour and hence, TPD is also known as brown blast. The outer bark gets dried and cracks open. Tumors can be seen on the panel area. Otherwise, the tree continues to grow normally. The exact cause of this syndrome is unknown although excessive harvesting of latex seems to be a possible trigger and high yielding clones are more vulnerable.

Management

Low frequency tapping is also recommended which will reduce the incidence of tapping panel dryness.

❑❑❑

4

Coffee

a) Die-back

Die-back is a physiological disorder that refers to the death of young branches during dry period. Pre-mature yellowing of leaves and drying of young tertiary branches from third and fourth node progressing either ways towards tip as well as downwards are its symptoms. It occurs due to adverse environmental factors, such as high temperature, high light intensity, low soil moisture, low relative humidity and depletion of reserve carbohydrates. Symptoms are interveinal chlorosis and narrow crinkled, small leaf formation with abnormal branching.

Management

Remove dead and whippy wood. Provide optimum shade (30%). Spray pre-blossoms and per-monsoon bordeaux mixture (0.5%). Conserve soil moisture status by thick mulch. Apply adequate quantity of fertilizer and nutrients. Liming of soils to correct pH is also recommended.

b) Premature Fruit drop

Premature fruit drop usually occurs in coffee during the berry development stage, viz., 90-120 days after blossom. The extent of fruit drop varies depending upon the field conditions (waterlogging-wet feet) as well as physiological status of the bush (hormonal imbalance, carbohydrate content and nutrient deficiency). 'Wet feet' condition in the plantations is due to continuous heavy rainfall with cool ambient temperature, high relative humidity, cloudy weather and improper drainage leading to waterlogging.

Management

The fruit drop can be checked by adopting proper soil cultivation methods like providing good drainage, cradle pits, etc. In addition, post-blossom (15 days after blossom) and per-monsoon application of one of these growth regulators (hormonal or Planofix or Miraculan or Atonik @ 50ml/200 litres of water or Potozome or Cytozyme crop plus @ 60ml/200 litres of water) is also useful in controlling premature fruit drop.

c) Bean Disorders

Bean disorders like black bean, black joloo and normal joloo are generally observed in arabica coffee due to physiological factors during bean development stage. It is reported to the extent of 2-10% from different localities. The main reason attributed to the bean disorders is the short supply or depletion of carbohydrate reserves during bean development stage. The depletion of plant reserves may be due to defoliation associated with rust disease and consecutive high crops, low light intensity because of overcast/hanging mist, hailstorm damage, soil saturation 14-16 weeks after flowering, skipping/ reducing fertilizer dosage, lack of proper bush management and maintenance of high shade.

Management

- Two rounds of bordeaux mixture as pre-monsoon and post-monsoon application (during May and September) is quite good to keep it under control.
- The post-monsoon application of bordeaux mixture could be given along with urea (0.5%) and muriate of potash (0.375%) (1 kg of urea and 750g of KCl/200 litres).
- Soil application of NPK at recommended doses and splits is recommended.
- Proper bush handling, shade management and other cultural operations are essential.

❑❑❑

5

Tea

a) Banjhi

Temperate woody plants like tea annually cycle through a period of active growth and a period of growth cessation and dormancy. The tender shoots with growing tips and young foliage, which would in normal course have contributed to food reserves, are perpetually harvested at short, regular intervals, stimulating a rapid succession of new crop of shoots in tea. The terminal bud of a growing shoot on a frequently harvested tea bush produces several leaves before it ceases leaf production and becomes dormant. Wangi or Banjhi stage of bud dormancy in tea is defined as a state in which leaf growth stop temporarily and the apical or lateral buds do not elongate, even when kept under optimum growing conditions due to mechanical/harvesting stress. This phenomenon occurs, when the leaf opening speed is higher than the formation speed of new leaf primordia. Tea is generally manufactured from two leaves and a bud which is obtained by plucking. There are two types of bud dormancy observed in tea:

i) winter dormancy is due to climatic condition and

ii) banjhi dormancy is due to mechanical stress.

It has been reported to be the result of either hormonal imbalance or lack of nutrient supply to actively growing apices. It could be easily differentiated from an active terminal bud as the former is very small compared to the latter. The dormant apical bud re-activates its leaf production after the period of dormancy which could be several weeks or months depending on various ecological and genetic factors. This characteristic of alternative active and dormant phases of growth of a free growing shoot is known as periodic growth or growth periodicity of tea. Accordingly, a tea bush in plucking has both actively-growing and dormant shoots. The 'banjhi' or dormancy cycle in free-growing tea trees appears to be the result of an endogenous rhythm. However, in harvested tea, apical dominance causes dormant shoots to be produced under conditions favourable to growth and the synchronized flushing cycle causes periods of apical shoot (active terminal bud) and secondary shoot (dormant terminal bud) production.

In tea, under cultural conditions of south India, such dormant buds develop throughout the year due to different abiotic stresses. Removal of young and growing shoots comprising the apical bud and the two internodes immediately below it, which constitute the crop in tea, is called plucking. Banjhi formation will have a direct impact on shoot initiation rate and shoot extension rate, which will ultimately, lead to reduced length, weight and yield. 'Banjhi buds' are not favoured since their presence has a detrimental effect upon the yield and quality of made tea.

❑❑❑

6

Cardamom

a) Hail Storm Injury

Cardamom are susceptible to hailstorm. The damage caused by hailstorm on plant tissue depends mainly on the size of the hailstones, duration of the storm event and the condition of the plant tissue, when the injury occurs. Hailstone damage to foliage and pseudostem appears as bruising and physical mangling. Large cardamom being a broad-leaved plant, the lamina tears parallel to the veins and symptoms on damage on leaves appear after a few days of the hailstorm.

The pseudo stem showed open and ragged-edged wounds that are produced on the vertically or obliquely oriented pseudo stems, depending on the direction of wind during the hailstorm. Subsequently resulted in partial or complete drying of the damaged foliage. This can cause the blight pathogen to invade through the wounds in the leaves and pseudo-stem and it leads to the blight disease.

Extensive damage could be seen on leaves in the exposed areas without any shade trees. It was observed that the plants in the areas protected with agro- shade net and those under shade trees having spread out canopies had no damage at all. The yield of the plant may be affected in the subsequent crop season when the floral parts of the plant are affected by the hail.

Management

The pandal erected in open places with shade net during March to May will help to manage the hailstorm damage to some extent and also helps to provide shade during summer. After the hailstorm, debris should be cleaned off and the broken pseudo stems and leaves should be trimmed off.

The application of organic manure like farmyard manure (2-5 kg) and vermicompost (1-2 kg) depending on the size of the clump can be taken upto the affected plants after the hailstorm which would help the plants form new foliage. Sufficient moisture should be maintained in the soil after the application of manures. The application of fungicides may also be undertaken to prevent the entry of the pathogen through the wounds.

b) Drenching of Fertilizer Solution

Experiment conducted at Indian cardamom research institute, Myladumpara, Idukki (Dt.), Kerala revealed that drenching fertilizer solution around the cardamom plant increased the yield as well as fertilizer use efficiency. The dosage used per plant for drenching of DAP, MOP and Urea combination are 30g, 50g and 20g respectively (dissolved in 5 litres of water). The fertilizer solution should be drenched in the plant base covering panicles. Fertilizer drenching in the above dose should be repeated nine times at monthly intervals avoiding heavy rainfall months of the year. (810 kg/ha) compared with soil application of fertilizers in three splits (445 kg/ha).

❑❑❑

7

Coriander

a) Frost Damage

Frost is the solid deposition of water vapour from humid air. It is formed, when the temperature of a solid surface is below the freezing point of water and also below the frost point.

Types of Frost

- Hoar frost
- Advection frost
- Window frost
- White frost
- Rime

The coriander crop is most vulnerable to frost damage at the flowering and early seed formation stages. The incidence can be minimizing by adopting the following control measures.

Symptoms of frost damage

- Scorching and pale brown patches forms between the leaf veins.
- Leaves scorched and turn brown and may eventually lead to the death of the plant.
- Collapsing of stems.
- Spring frosts can damage blossom and young fruits.
- Late spring frosts can cause leaf scorch, browning and even total plant death.
- Spotting on the leaves.

- An early symptom of frost damage appears as water-soaked and dark-green, turning black in time.

Management

- Spray 0.1% solution of sulphuric acid.
- Irrigate crop just before the incidence of frost.
- Set up wind breaks obstructing the cool waves.
- Create smoke cover in the early morning to dilute the effect of cold waves.
- Spraying of proprietary chemical to limit potential frost damage.
- Installation of overhead irrigation for protecting the plants.
- Introduction of Russian Cultivars in coriander which are having cold resistance.
- Cover plants with a double layer of horticultural fleece or other suitable protection when frost is forecast.
- Select planting positions carefully to avoid 'frost pockets'.
- Choose plants that are reliably hardy and suited to your growing conditions.
- Developing varieties resistant to frost damage.
- Soil loosening should be avoided, whenever frost is expected.

V. Flower Crops

1

Greenhouse Roses

a) Blindness

Flower formation in greenhouse roses is a self induced function and initiation of flowers begins in an actively growing meristems within 14-16 days after the removal of the previous marketable stem. By this time, the newly emerged axillary shoot should be about 5 cm long. Each of these developing shoots would normally terminate in a flower but under some conditions, the flower buds may :

- Fail to form at all
- Begin to form and then abort along with the uppermost few leaves, or
- Atrophy and abscise. Whatever occurs, the end results is a "blind shoot".

These remain short and then they elongate very slowly and only bear a few leaves. Lower axillary buds have a greater tendency to be blind. Blindness is more likely to occur during periods of low light and temperature than under brighter warmer conditions since temperature affects the concentration of plant hormones. If the production of such growth-promoting hormones is low, then the initiating flower bud is likely to abort during an early stage of development.

Blindness in roses appears to be related to a number of critical metabolic processes going on within the plant, the most influential being competition for a limited supply of photosynthates and other metabolites. This competition is affected by apical dominance favours the upper shoot on a stem where more than the upper most bud has broken. Gibberellin are assumed to encourage flower development by causing the developing bud to become a more powerful metabolic sink for the raw materials.

b) Bullheads

This malformation is characterized by :

- A lower than normal ratio of length to diameter of the flower bud, giving it a flat-topped appearance.
- An increase in size and weight of the flower bud.
- An increase in the number of short petals and petaloids
- The proliferation of secondary florets hearing mainly carpels near the base of the flower.

Bullheads are more prevalent when temperature drop to 12-15°C during early flower development and are correlated with marked charges in the balance of different growth regulators present within the flowers. Low temperatures appear to reduce gibberellin activity and increase cytokinin activity which promotes the development of the adventitious florets which are associated with bullheads.

c) Bent Necks

When the cells in the neck area lose a turgor, the neck bent or droops due to lack of mechanical strength. This is essentially a water balance problem in which the rate of water loss from the roses exceeds the rate of water uptake. This causes a lose of turgor of the cells in the pedicel just below the flower. A common cause of bent neck is an impeded flow of water through the stem of the cut rose due to air blockage of the water-conducting vessels, physiological plugging or direct or indirect microbial plugging of the water conducting system. Post-harvest handling procedures have been shown to help prevent the problem, including recutting rose stem under water and using solutions containing antimicrobial chemicals. Another common cause of bent necks is premature harvesting of buds.

❐❐❐

2

Jasmine

a) Chlorosis

The foliage will exhibit interveinal chlorosis as a result of iron deficiency. When the plant is infested with nematode, the iron and other nutrient availability by translocation to the plant tissues is hampered due to blockage of vascular bundles as a result, it is made to suffer from shortage of iron and other nutrients as also water. The leaves will have fade yellow colouration; the growth of the plant is stunted which may eventually lead to the wilting of the entire plant. The flower yield will be substantially reduced.

Management

After digging the soil around the tree, the following are to be applied to the soil: 10g of Temic or 20g of Furadon and 250g of Neem cake. It is followed by irrigation and application of 0.1% emison solution. After one week, foliar application of 500 g of ferrous sulphate and 100 g of urea along with 100 ml of soap solution mixed in 100 lit of water is done two to three times at 20 days interval depending upon the extent of severity.

❑❑❑

3

Chrysanthemum

a) Plants too Short

Plants that are too short can be caused by several factors, such as poor root growth, insufficient nitrogen in the early growing stages, failure to provide enough long days, or excessive use of growth retardants.

b) Plants too Tall

Excessive height can be caused by too many long days, crowded conditions, inadequate light intensity and temperatures that are too high. A high day temperature combined with a low night temperature can cause plants to grow much taller.

c) Uneven Flowering

Uneven flowering may be a problem, especially in winter, because of cool nights, as flower buds fail to form at night temperatures below 60°F. Stray lights striking plants during desired dark periods can also cause this problem.

d) Inadequate Shoots

The development of too few shoots usually occur during the first two weeks after plants are pinched. Some problems that may develop during this time that could cause fewer shoots to develop are poor root growth, cool night temperatures, very dry air and insufficient nitrogen.

e) Malformed Flowers

Malformed flowers can be caused by diseases or insects, but poor control of day length, temperature, or both can also affect flower form.

f) Failure to Flower

Ethylene pollution from faulty heaters, incorrect day length manipulation, light leaks when shading and extreme temperatures are some factors that can cause this disorder.

g) Poor Growth

Poor growth can be caused by not enough sunlight in very dark weather, too much or too little fertilizer, improper pH, overwatering and poor drainage. Farmers need to keep records of their cultural practices and also some notes on extreme weather conditions. These conditions will help to determine the causes of some poor growth problems.

h) Wilting

Plants will often wilt if the weather has been overcast for a few days followed by bright sunny days. This results from the plant not being able to absorb water quickly enough. This condition should correct itself after a few bright days.

i) Crown Buds

The major differnce between crown buds and terminal buds is that leaves beneath a crown bud are strap-shaped, while leaves beneath a normal terminal bud are lobed. Vegetative shoots can emerge around a crown bud, if the causative factor is not corrected.

j) Bract Buds

During the warm summer months, it is not unusual to find overdeveloped individual floret bracts on some cultivars. Bract buds may develop, when day or night temperatures exceed 81°F.

k) Sunscald

Sunscald develops, when high temperatures and high light intensities cause rapid evaporation of moisture from flower petals and is often confused with *Botrytis blight*. Sunscald and *Botrytis blight* are easy to distinguish because the former occurs on younger petals at the centre of the flower and the latter infests the tips of older petals. ***Botrytis*** will often affect petals damaged by sun scald. However,

Sunscald can be prevented by supplying a light shade over the plants as colour starts to show on the flower buds and by proper selection of cultivars for certain areas or time of the year.

l) Faded Flowers

Plants exposed to cool temperatures (below 60°F) or flowers that are old will have a pinkish hue on white flowers. High temperatures, above 75°F, will often cause fading in dark coloured flowers or failure of the floral pigments to develop.

m) Air Pollutants

Leaf discolouration is sometimes caused by air pollutants. stunted growth, malformed flowers and delayed flowering have often been attributed to air pollutants.

n) Quiling of Florets

Petals get twisted and become cup-shaped. The peripheral flowers lose it's turgidity. This is due to boron deficiency and low night temperature (< 15°C).

Management

Boric acid sprayed @ 3g/litre during vegetative, bud initiation and flowering stage.

o) Pink Colouration of White Petals

Night temperatures below 15°C causes pink colour in white petals.

p) Delayed Flowering or no Flower bud Formation

Bud development is completely arrested and the formed buds develop slowly and stop development or abort. This is due to too long photoperiods and low night temperature and also due to copper and manganese deficiency.

Management

- Application of Manganese sulphate (3g/litre) applied at vegetative, bud initiation and at flowering stage.
- Application of Copper sulphate (1 g/litre).

q) Yellowing

It is characterized by breakdown of chlorophylls, proteins and nucleic acids in the detached leaves which accentuates, when the cut flowers are held in dark at high temperature. Foliar yellowing is a major problem in chrysanthemum cultivars and caused by many factors, namely :

- High temperatures which encourage the process of senescence including leaf yellowing in all the flowers.
- Light significantly reduces senescence and leaf yellowing.
- High levels of nutrients particularly nitrogen during growing period encouraged post harvest yellowing in chrysanthemum.

Management

Cytokinins are known to delay senescence of leaves. A foliar spray with the cytokinins BA @ 20 to 100 ppm before shipment or storage greatly delayed yellowing and maintained the turgidity of leaves. Dipping of foliage is more effective than spray even at a lower concentration.

❏❏❏

4

Gladiolus

a) Geotropic Bending of Spikes

The tips of gladiolus spikes show tendency to bend against gravity if placed horizontally for longer periods. It is due to the accumulation of IAA on the lower portion of spike that causes asymmetrical elongation of cells.

b) Blindness

It results in complete absence of spikes. It is due to the deficiency of zinc.

Management

- Foliar application of Zinc @ 20-100 ppm.

c) Topple

The collapse of a small portion of the internode just beneath the flower is called sugar stem or wet stem and it is due to the deficiency of calcium.

Management

Calcium nitrate @ 24 ppm can be sprayed.

❑❑❑

5

Gerbera

a) Bent Neck

It is the insufficient flower stem hardening or insufficient maturation of the stem tissues below the harvested flower which results in stem collapse and drying of plants too frequently. The main causes are related to bacterial contamination which is sensitive to ethylene.

b) Double Faced Gerbera

Vegetative growth is very high while the flower buds are small. It is due to the imbalance of nutrients.

c) Scape Bending

This phenomenon is caused by excessive bacterial growth in vase solution which reduces longevity.

d) Premature Flower Wilt

Premature wilting of flower occurs while stems are still attached to the plants and often develops just as petals are in full expansion. Lack of storage carbohydrates is the main reason for this problem.

e) Pre Harvest Stem Break

It is common post-harvest disorder in cut gerberas. The main causes are water imbalance and early senescence associated with water stress.

❑❑❑

6

Anthurium

a) Colour Breakdown of Spathe

This can be recognized by colour breaks, which may or may not be accompanied by distortion of the spathe. The suspected cause is genetic mutation within the spathe. It may also be due to calcium deficiency.

Management

- Application of Calcium nitrate @ 24.75g/m^2.
- Lime should be applied @ 5g/plant/month.

b) Sticking

Flowers do not open properly and spathe is stuck. It is variety dependent. Low relative humidity is the main cause.

c) Jamming

Flowers get jam in the sheath. The leaf is wound too tightly around the flower. It is variety dependent and occurs during arid conditions. It also occurs more frequently in cultivars with long sheaths. Humidifying the top layer of the substrate is recommended to provide a favourable micro-climate.

d) Folded-ears

The suspected cause is an environmental-physiological interaction. Temperature below 15°C causes this disorder.

e) Sunburn

Overall fading of the spathe colour or browning of the spadix which is caused by direct sunlight, light shade levels and flower age. Largest flowers show normal colour.

f) Crooked Stem

The suspected cause is an environmental-physiological interaction involving excessive nitrogen feeding and fluctuating wet and dry condition. Heavy shade levels and lack of pruning may also contribute to crooked stems.

g) Bleach

In mild cases, impaired colour development occurs in the lobe area of the spathe and in severe cases, the entire flower including the spadix may show the signs of insufficient colour development, stunting, distortion and necrosis. The suspected cause of bleaching is an environmental-physiological interaction involving excessive application of ammonium nitrogen, excessive salt build up and fluctuating low and high temperatures.

h) Flower Abortion, Flower Deformation and Rosette Formation

Flower spikes aborts and growth of the spathe is stunted and the stem of the flower is also short. This occurs under conditions in which the plant builds up a strong pressure that cannot be processed by the plant above ground.

□□□

7

Carnation

a) Calyx Splitting

Sepals beneath the flower are unable to form a cylindrical calyx tube which supports the base of the petals. As the bud opens and size of the petals increase, calyx may split down either half or completely, because it is unable to contain the load of extra petals. Many factors are responsible for calyx splitting viz., boron deficiency, irregular or fluctuating temperature during flowering and low night temperature below 10°C.

Management

- Selection of resistant varieties.
- Regulation of temperature-day temperature (20-25°C), night temperature (12.5-15.5°C).
- Application of nitrogen @ 25-40 ppm and boron @ 20-25 ppm.

b) Calyx Tip Dieback

Browning begins at the calyx tip and progresses downwards until much of the calyx is damaged. Water stress and potassium deficiency are the major cause for this problem.

Management

- Application of KCl @ 5 g/litre of water (two sprays at 10 days interval).
- Provide adequate irrigation @ 4.5 litre/m^2.

c) Tip Burn

Young leaves do not open evenly and the leaves at the tip curls followed by drying. Calcium deficiency is the main cause for this problem.

Management

- Application of Calcium nitrate @ 5 g/litre (10 days interval).

d) Sleepiness

Sleepiness in carnation is caused by exposure of the flowers to ethylene gas or water stress. The extent of this disorder is much higher, when flowers are stored for long periods or when they are kept at high temperature.

Management

- Provide adequate irrigation @ 4.5litre of water /m^2.

e) Grassiness

Grassiness refers to failure of plants to produce flowers. This is a genetic disorder which varies from variety to variety. Removal and destruction of affected plants is the only way of correcting this disorder.

f) Slab Side

This disorder refers to uneven opening of flower buds resulting in the petals protruding on one side only, giving an asymmetrical and lopsided shape to the flower. It is common during cooler periods.

Management

This can be overcome by gradually increasing the temperature to optimum level.

g) Internode Splitting

Splitting of internodes affects the quality of cut flowers. It is due to boron deficiency.

❑❑❑

8

Azalea

a) Failure to Flower

Late pinching, improper temperatures and excessively low light intensity during the flower initiation and early developmental stages are the main cause for this problem.

b) Defoliation

Defoliation can be a serious problem and there are several reasons why evergreen azalea plants could lose their leaves. Almost any type of stress could be responsible, such as lack of water, fertilizer injury, disease and ethylene pollution. Leaf abscission caused by ethylene is most likely to occur in a cold storage until that might have some ethylene-producing material in it. Inadequate lighting (intensity or duration, or both) in a cooler, when the temperature is over 40°F will almost always cause leaf abscission. Low relative humidity, lack of water in the medium and excessive air movement can also cause the foliage to abscise. Leaf burn will precede abscission in some instances, such as excessive fertilizer or diseases, but green leaves can abscise if the causes are ethylene, low light, or lack of moisture.

c) Sparse or Irregular Flowering

Adequate time from the final pinch to the dormancy breaking treatments must be allowed to enable shoot apexes to complete the reactions that are necessary for uniform flowering to occur. Flower bud initiation will be delayed and further development will be inhibited, if plants are pinched too late in the season. Uneven flowering can also occur if shoot apexes on a plant are of different ages. Shoots that were present but too small to pinch will often form flower buds sooner than lateral shoots that evolved from the pinch. Those flower buds could also force faster after the cold storage treatment. Excessive crowding of plants during the vegetative and early reproductive stages could also result in the flowers in the upper part of the canopy opening sooner than those on the sides of the plant, where light is reduced.

d) Herbicide Injury

Twisted leaf petioles, leaf burn or spotting and many other symptoms could reveal contamination by an herbicide. Herbicide use in the greenhouse is really restricted and only chemicals labelled for greenhouse use should be applied.

❏❏❏

9

Lily

a) Leaf Scorch

The cause is high levels of fluoride and perhaps other elements in plants. Leaf tips or crescent-shaped segments of leaf margins turn yellow and die. Major sources of fluoride are fluorophosphates (e.g., superphosphate), or irrigation water.

b) Abnormal Leaves and Flowers

Abnormal growth patterns, leaf distortion, or both may occur because of virus infection, nutritional deficiencies or toxicities, high soluble salts, exposure to herbicides and abnormal temperatures. Aphid infestations can cause malformed flowers.

c) Bud Abortion and "Blasting."

Death of young or well-developed flower buds has been attributed to the following causes: root injury causing water stress, high soluble salts, or root rots when flower buds are visible; low carbohydrate supply; inadequate light intensity caused by shading or too close plant spacing, especially when plants are being forced at high temperatures and application of growth retardant. The problem also seems to be related to ethylene as low as 1 ppm for 2 days in the greenhouse.

d) Yellow Lower Leaves

Root system damage from disease or high soluble salts will result in yellow, perhaps wilted and drying lower leaves. Similar symptoms are caused by excessive rates of chemical growth retardants, especially when coupled with low phosphorus levels in plant leaves. Low mineral nutrient levels in plants caused by termination of fertilizer application two to four weeks before marketing also promotes yellowing of leaves.

❑❑❑

10

Poinsettias

a) Leaf Distortion

The symptoms are extremely variable. In some cases, damage has occurred only at the tip of the immature leaf and it will give the appearance of having been chopped off at a later stage of development. Where the entire margin of the leaf has been affected in earlier stages, later growth of all except the margin may cause a "puckered" appearance. It seems that when cells in very young leaf tissues are ruptured or killed, the leaf becomes misshapen as it expands. Drying of tissue, burn from fertilizer, nutrient deficiency (particularly molybdenum) and chemical burn have all been suspected of damaging young leaf cells. Plants under stress from bright light, extremely warm temperatures, or moving air often have more leaf distortion.

b) Bract Edge Burn

Severe bract burn has been encountered, where extreme rates of fertilizer have been used. Under these conditions, the leaves may show no damage. One theory is that during growth, there is a diluting effect of plant-absorbed fertilizer, but at flowering, new tissue development has virtually ceased and the fertilizer nutrients accumulate in the youngest mature and most sensitive tissue - the bract. This accumulation causes cell damage, usually starting on the bract edges. Where slow-release fertilizers are used, the rate should be modest and the application should be early enough to ensure almost complete depletion at time of flowering. It may also help to increase the intensity of irrigation in the finishing stages to ensure adequate leaching and removal of accumulated salts.

Bract burn may also be triggered by environmental factors, namely water stress, temperature, light intensity and humidity.

c) Premature Cyathia Drop

The true flowers, or cyathia, may drop from the center of the bract presentation before the flowers reach maturity. The premature cyathia drop is caused by low light levels, high forcing temperatures, or both. Water stress aggravates the problem. These conditions allow the food reserves of the plant to become depleted. As the food reserves become low, the plant reacts by dropping the cyathia. It is not uncommon for skies to become overcast during late October and November in northern states. Lower light levels and cooler temperatures accompany the cloudy weather. These are not ideal conditions for high quality poinsettias. With lower light levels, plants do not make as much food. Also, as poinsettia bracts develop, they shed the green leaves below and further restrict the amount of light available for photosynthesis. If poinsettias are grown at lower than optimum temperatures during the early part of the production period, flower development may be delayed, when the cloudy weather begins. To speed-up the rate of flower development, it then becomes necessary to raise greenhouse temperatures. As temperatures increase, the plant's food reserves are used at a faster rate, increasing the possibility of cyathia drop.

d) Latex Eruption

Plants belonging to the Euphorbiaceae family contain latex, which is exuded when cells are injured. The mechanism is one of the bursting cells resulting from high turgor pressure with latex spilling over the tissue and on drying, the creation of a growth restricting layer. When this occurs on developing stem tips, stunting of growth occurs. The exuding of latex has also been observed on fully expanded leaves, sometimes giving the appearance of mealy bug infestation because of the white splotches scattered over the leaf surface.

All of the contributing factors have not been clearly defined, but several obvious ones include high moisture availability and high humidity, both of which result in high fluid pressure within the cells. Low temperature is an important contributing factor. Mechanical injury from rough handling or from excessively vigorous air movement may also increase injury to cells. High rates of photosynthesis may contribute by building up a high osmotic pressure in cells from carbohydrate accumulation. Control is best attained by using growing media that dry out in a reasonable length of time and by avoiding extremes of high humidity, particularly during the night. Moderate shading in extremely bright weather might also be helpful.

e) Stem Splitting

Under certain conditions, poinsettias that do not normally branch unless pinched will suddenly produce stem branches at the growing tip. Careful examination reveals that the true stem tip has stopped growth or aborted. Splitting is actually a first step in flower initiation. The stimulus to flower increases with the age of the stem and with the lengthening of night. Even with short nights and normal 60° to 70°F temperatures, splitting can be expected if the stem is permitted to grow until 20 to 30 leaves are present. Stem tips that are continuously propagated have an increasing tendency to flower. It is always good insurance to discard early pinched shoots instead of trying to propagate them. Also, stems that are heavily shaded by a canopy of higher foliage may be subjected to reduced light to cause them to split even in periods, when day length would be considered adequately along to keep apexes vegetative.

f) Bilateral Bract Spots

This condition is characterized by breakdown of the tissues between the veins located on either side of the bract midrib. It occurs in late November and early December during the flowering process. Although the plant is not killed, the condition can effectively lower the quality of the plant, rendering it unsalable. Bilateral bract spots seem to occur under a wide temperature range in various types of greenhouses.

❑❑❑

References

Abeywickrama, G. and Weerasingee, S.S.2002. Physiological disorders of Banana in the southern dry region. Annals of the Sri Lanka Department of Agriculture.

Adams P, 1993. Crop nutrition in hydroponics. Acta Hort. 323: 280-305.

Ahlawat, T.R., Singh, A., Patel, C.R. and Rymbal, H. 2014. Management of physiological disorders in mango. Indian Farmer 1(3): 1285-197b.

Bal, J.S. 2006. Fruit growing. Kalyani Publishers, Ludhiana.

Barche. S and Kirad, R.K. 2009. Fruit Cracking - A New Emerging Problem of Guava. Indian Farmers' Digest, (November 2009) p. 33.

Basak, R.S. 2005. Malnutrition and Treatment of Vegetable Crops. Kalyani Publishers, New Delhi.

Bhat, R., Sujatha, S., Bhavishya and Karthika, K.S. 2016. Nutrient Management for Sustained Productivity of Arecanut. Indian J. Arecanut, spices and medicinal plants 18 (4): 26-28.

Bruton, B.D. 1994. Mechanical injury and latent infections leading to postharvest decay. Hort.sci. 29: 747-748.

Cakmak, I. 2002. Plant nutrition research: Priorities to meet human needs for food in sustainable ways. Pl.Soil, 247: 3-24.

Carne, W.M. 1948. The non-parasitic disorders of apple fruits in Australia. CSIR Bull. No. 238.p. 83.

Ceponis M.J, Cappellini, R.A, Lightner, and G.W. 1987. Disorder in cabbage, bunched broccoli and cauliflower shipments to the New York market. Plant Dis., 71: 1151-1154.

Christensen J.V. 1994. Causes of fruit cracking. In: Proceedings of International Conference on Cracking in Cherries. Department of Horticulture, Michigan State University, East Lansing, Michigan, USA, p.7.

Chundawat, B.S., Jindal, P.C. and Khera, A.P. 1979. Effect of girdling on bud, flower and berry drop in grapes. Haryana Agric. Univ. J. Res. 9:332-336.

Chutinunthakun. T. 2001. Prevention of the incidence of translucent flesh disorder and internal gumming fruits in mangosteen (Garcinia mangostana Linn.) and screening techniques. M.Sc. Thesis. Prince of Songkla University, Thailand.

De, L.C. and Bhattacharjee, S.K. 2011. Handbook of Vegetable Crops. Pointer Publishers, Jaipur (Rajasthan), India.

Devaraju, T., Srinivasa, V., Ravi, C.S. and Kantharaju, V. 2013. Physiological disorders in Tomato and their Management. Plant Hort. Tech. (April-May 2013). pp 30-31.

Dhanapal, K., Gudade, B.A., Bora, S.S. and Gautam, A and Rema Shree, A.B. 2017. Hailstorm damage in large cardamom. Spice India English, (August 2017), pp. 25-26.

Dhillon, W.S. 2013. Fruit Production in India. Narendra Publishing House, Delhi.

Dorais, M., Demers, D.A., Papadopoulos, A.P. and Van-Ieperen, W., 2004. Greenhouse tomato fruit cuticle cracking. Hort. Rev, 30: 163-184.

Edward Raja, M. 2009. Importance of micronutrients in the changing horticultural scenario in India. J. Horti. Sci. 4: 1-27.

Elyatem, S. M. and Kader, A. A. 1984. *Post-harvest physiology and storage behavior of pomegranate fruits*. Scientia Hort. 24:287-298.

Fernandez-Trujillo, J.P., Cano, A., Artes, F., 1998. Physiological changes in peaches related to chilling injury and ripening. Postharvest Biol. Tech. 13; 109-119.

Gajbhiye, R.P. 2012. Keeping tomato free from disorders. Indian Hort.(January-February 2012) pp. 8-9.

Ganai, N.A. 2015. Managing physiological disorders in apple for better return. Indian Hort. (November-December) pp. 24-27.

Gardner, C.M.K., Laryea, K.B. and Unger, P.W. 1999. Soil Physical Constraints to Plant Growth and Crop Production. Land and Water Development Division, Food and Agriculture Organization, Rome.

Gorguet, B., Heusden, V. A.W. and Lindhout, P., 2005. A review on parthenocarpic fruit development in tomato. Plant Bio. pp: 131-139.

Gowda, J.V.N. 1999. Physiological Problems in Greenhouse Roses, Floriculture Today (April 1999), pp 15-16.

Gunasekaran, M. and Chandramohanan, R. Crown Choke of Coconut. In: Gangawane, L.V. and Khilare, V.C. 2008. Crops diseases, Identification and Management - A Colour Handbook, Daya Publishing House, Delhi.

Gunasekhar, N. and Muthukumar, V.B. 2001. Grapes: Nutritional and Physiological Disorders. Kissan World (May 2001). pp 57-58.

Gupta, U.C. 1979. Boron nutrition of crops. Adv. Agron. 31, 273-307.

Gupta, V.K and Paul, Y.S. 2005 Diseases of plantation crops. Kalyani Publishers, New Delhi. pp46-47.

Gupta, S.K. 2015. Reclamation and Management of water-logged saline soils. Agric. Res.J. 52: 104-115.

Goswami, R.G. and Bajpai, R.K. 2017. Managing chemically deteriorated soils of India. Indian fmg. 67: 4-6.

Harender, R., Bhardwaj, M.L. and Raj, H., 1995. Sunscald damage in summer season tomato. Indian Phytopath, 48(3): 367-368.

Hazra, P. and Som, M.G. 1999. Technology for Vegetable Production and Improvement. Naya Prakash Publications, Calcutta. p. 148.

Heena, M.S. 2015. Physiological disorders in root crops and their management. Plant. Hort. Tech. (September 2015) pp. 52-55.

Hoftun, H., 1993. Internal atmosphere and watery scales in onion bulbs. Acta Hort. 343: 135-140.

Ingle, M. and D'Souza, M. C. 1989. Physiology and control of superficial scald of apples: a review. Hort Sci. 24:28-31.

Jalikop S. H., Venugopalan, R and Kumar, R. 2010. Association of fruit traits and aril browning in pomegranate. Euphytica 174:137–141.

Joshi, P. 2008. Fruit Drop in Citrus: Causes and Control Measure. Indian Farmer's Digest (July 2008) pp. 24-26.

Ju, Z., Duan, Y., Ju, Z., 2000. Leatheriness and mealiness of peaches in relation to fruit maturity and storage temperature. J. Hort. Sci. Biotech. 75, 86-91.

Kannan, M., Jawaharlal, M. and Ranchana, P. 2012. Physiological disorder which hamper cutflower production. Floriculture Today (May). pp 30-32.

Karunanithi, K., Karthikeyan, M., Sarala, L., Mohandas, S. and Rajarethinam, S. 2002. Button Shedding in Coconut. Indian Coconut J. (April) pp 10-11.

Kennelly, M., O'Mara, J., Rivard, C., Miller, G.L. and Smith, D. 2012. Introduction to Abiotic Disorders in Plants. Plant health instructor 10: 10-29.

Kolekar, H.D., Alloli, T.B. and Asundi, J.A. 2001. Irregular bearing in Mango. Kissan world (October) pp. 26-27.

Kumar, A. and Mohanta, R.K. 2018. Preserving soil health for sustainable production. Kurukshetra (February 2018) pp. 29-31.

Kumar .A., Singh .C., Ral .M. and Ranjan R. 2001. Effect of irrigation, calcium and boron on fruit cracking in litchi cv Shahi. Orissa J. Horti. 29: 55-7.

Kumar, M.K. and Rani, M.U. 2013. Frost damage and its control in spice crops. Indian J. Arecanut, Spices & Medicinal Plants, 15(3): 13-15.

Kumar, R. 2015. Rejuvenating old, senile orchards of tropical and subtropical fruits for enhanced production and imporved quality: A review. Indian J. Agri. Sci. 85: 295-313.

Lal Saran, P., Choudhary, R and Solanki, I.S. 2017. Fruit deformity in papaya: field screening, nutrient composition and amelioration by boron application. Indian J. Exp. Bio. 55, pp. 870-875.

Larson, R.A. 1992. Introduction to Floriculture, 2nd edition Academic Press, London.

Manimaran, P., Priyanka, R., Ghosh, S. and Alen, J. ,20,17. Major physiological disorders of temperate fruit crops. Agrobios Newsletter, 16. 57-58.

Marboh, E.S., Singh, S.K., Pandey, S., Nath, V., Gupta, A.K. and Pongener, A. 2017. Fruit cracking in litchi. Indian J. Agri. Sci. 87: 03-11.

Matthais, J. P. 1995. Factors contributing to internal breakdown of Fuji apples. Washington State University Tree Fruit Postharvest J. 6:3-4.

Meheriuk, M., Prange, R.K., Lidster, P.D. and Porritt, S.W. 1999. Post-harvest disorders of apples and pears. Agriculture Canada Publication, Communications Branch Agriculture, Canada.

Mester, I.M. 1971. The result of investigations on the boron nutrition of grapevines. Hort. Abstr.43:5122.

Misra, A.K., Selvarajan, M., Subbiah, A. and Thamariselvi, S.P. 2013. Reaping mango fruits round-the-year. Indian Hort. (July-August 2013). pp. 3-7.

Mitra S. K, Dutta Ray S K and Mandal D. 2014. Control of fruit cracking and sun-burning in litchi by irrigation and moisture conservation. Acta Hort. 1024: 177-81.

Nancy, D., 2015. Scrutiny of strategies in developing parthenocarpic tomato. Indian J.Appl. Res., 5(3): 71-73.

Nandini, K.S., Archana, R. and Pallavi, N. 2017. Parthenocarpy in Tomato. Agrobios news letter: 16: 54-55.

Nelson, S.C.2008. Lightning Injury. Cooperative Extension Service, College of Tropical Agriculture and Human Resources, University of Hawaii at Manoa. p.1-2.

Oommen, M., Nagarajan, P. and Varghese, J.J. 2015. Drenching: a better way of fertilizer application in Cardamom. Spice India (February). 20-21.

Pal, R.N., and Chaddha, K.L. 1980. Black-tip disorder of mango: a review. Punjab Hort. J. 20:112-121.

Pathania, N.S. 1998. Specific Problems in Hybrid Lilies Cultivation. Floriculture Today, (June) pp 33-35.

Peirce, L.C. 1987. Vegetables – Characteristics, Production, and Marketing. John Wiley & Sons, New York.

Perring, M.A.; Pearson, K. 1986. Incidence of bitter pit in relation to the calcium distribution in the fruit. J. Sci. Food Agric. 37:709-718.

Prajna, S.P. 2015. Physiological disorders of root crops. Plant Hort. Tech. June 2015 pp.16-18.

Prakash, OM. 2003. Compendium of Mango Diseases and Disorders. Capital Publishing Company, New Delhi.

Prakash, G.S and Satisha, J. 2017. Revolutionizing grape industry. Indian Hort. (May-June (2017) pp. 26-28.

Prasad, R. and Shivay, Y.S. 2017. Sulphur fertilization and food quality - A review. Indian J. Agron. 62(1) 1-7.

Prasad, A., Saran. M.D. and Singh, K. 1971. Black-tip disease of mango and its control. Hort. Adv. 8:4-10.

Prince, T.A. and Cunningham, M.S. 1989. Production and storage factors influencing quality of potted Easter lilies. Hort. Sci. 24 (6). 982-984.

Purvis, A.C. 2004. Regulation of oxidative stress in horticultural crops. Hort.sci 39: 930-932.

Raese, J.T. 1989. Physiological disorders and maladies of pear fruit. Hort. Rev. 11:357-411.

Rajput, M.S., Kumar, J.S. and Bajwa, M.S. 1971. Control of black tip of mango with caustic soda. Punjab Hort. J. 11:49-51.

Ram, S. 1992. Naturally occurring hormones of mango and their role in growth and drop of fruit. Acta Hort. 321:400-411.

Ramteke, S.D. 2012. Physiological Disorders in Grapevine and its Management. National Research Centre for Grapes, Pune. p.35.

Rana, M.K. 2011. Physio-Biochemistry and Biotechnology of Vegetables. New India Publishing Agency, New Delhi.

Rane, D.A., Katrodia, J.S. and Kulkarni, D.N. 1976. Problem of spongy tissue development in mango (Mangifera indica L.) cv. Alphonso: A review. J. Maharashtra Agric. Univ. 1:89-94.

Rao, V.S. 2000. Principles of Weed science oxford and IBH Publishing Co. Pvt. Ltd., New Delhi.

Ravindra, V., and Shivashankar, S. 2006. Spongy tissue in Alphonso mango II: A key evidence for the causative role of seed. Curr. Sci. 91: 1712-1714.

Reddy, D.B., and Kapoor, S.P. 1965. Borax is the answer for mango black tip. Indian.fmg. 15:13-14.

Rizvi, S. J. H., Tahir, M., Rizvi, V. Kohli, R. K. and Ansari, A. 1999. "Allelopathic Interactions in Agroforestry Systems." Critical Reviews in Plant Sciences 18: 773–779.

Saikia, M., Gogoi, S. and Saikia, U.N., 1999. Management of blossom end rot of tomato by using different calcium sources. Ann. Biol. 15: 217-218.

Salaria, A.S. and Salaria, B.S. 2014. A to Z solutions fruits and plantation crops (Horticulture at a Glance Vol. I). Intellects-Nurture to Excel, Intellects Institute of Knowledge Management, Jammu.

Sandhu, S and Gill, B.S. 2013. Physiological disorders of fruit crops. New India Publishing Agency, New Delhi.

Saran, P.L., Kumar, R. and Johari, R.K. 2008. Fruit cracking disorder: An emerging problem of fruit crops. Indian Farmers. Dig. 41:30-32.

Sarvas D. and Ntatsi. G and Passam. H.C. 2008. Plant nutrition and physiological disorders in greenhouse grown tomato, pepper and eggplant. The European J. Plant Sci and Biotech, 2: 45-61.

Sdoodee, S and Limpun-Udom, S. 2002. Effect of excess water on the incidence of translucent flesh disorder in mangosteen (Garcinia mangostana L.). Acta Hort. 575:813-820

Semple, A., Thompson, A. 1988. Influence of the ripening environment on the development of finger drop in bananas. Sci. Food Agri. 46 (2): 139-146.

Sen, P. 2017. Micronutrients - Their availability and role in disease control mechanism. In: 82nd Annual convention of Indian society of Soil Science, New Delhi and National seminar on "Development in soil science - 2017, held on December 11-14, 2017 at Amit University, Calcutta. pp 121-135.

Shanmugasundaram, K.A., Thangaraj, T. and Muthuvel, I. 2007. Fruit cracking in Pomegranate and its management. Kissan world (June) p. 37.

Sharma, R.R and Shukla, R. 2005. Managing fruit pitting in mango. Indian Hort. (May) p. 28.

Sharma, R.R. 2006. Fruit production: Problems and solutions. Int. Book Distributing Co., Lucknow U.P., India.

Sharma, R.R., and Shukla, R. 2006. Managing fruit pitting in mango. Indian Hort. 51:28-29.

Sharma, R.R., and Singh, R. 2009. The fruit pitting disorder: A physiological anomaly in mango (*Mangifera indica* L.) due to deficiency of calcium and boron. Sci. Hort. 119:388-391.

Sharma, S., Sharma, R.K. and Yamdagni, R. 1997. Water berry development in Grapes - A review. Haryana J. hort. Sci. 26 (3-4) : 220-227.

Sharma. R.R. 2004. Granulation: A serious disorder of citrus. Intensive Agri. (January-February). pp 20-22.

Shivashankar, S. 2014. Physiological disorders of mango fruit. Hort. Rev. 42:313-347.

Shukla, A.K., Tiwari, P.K., Singh, P. and Tiwari, K.N. 2017. Macro importance of micronutrients in Indian Agriculture. Indian fmg. 66: 27-32.

Siddiq, M. 2017. Hand Book of Mango Fruit: Prtoduction, Post-Harvest Science, Processing Technology and Nutrition. John wiley and Sons.

Singh, D.B., Kingsly, A.R.P and Jain, R.K. 2006. Controlling fruit cracking in pomegranate. Indian Hort. (January-February) p. 14.

Sinha, S and Sharma, M. 2013. Physiological Plant Diseases. Campus Books International, New Delhi.

Smiley, E.T. 1907. Air Pollution Injury to Landscape Plants. Bartlett Tree Experts, Scientific Tree Care.

Subramanyam, H., Krishnamurthy, S., Subhadra, N.V., Dalal, V.B., Randhawa, G.S. and Chacko, E.K. 1971. Studies on internal breakdown, a physiological ripening disorder in Alphonso mangoes. Trop. Sci. 15:203-210.

Tewari, K.N. and Tiwari, R.2015. Managing boron deficiency in Vegetable crops. Indian Hort. (January - February) pp. 16-19.

Tomana, T. 1963. Histological and physiological studies on the cause of Jonathan spot in apples. Bull. Yamagata Univ. Agric. Sci. 4:1-63.

Usha, K., Dubey, A.K. and Singh, B. 2005. Citrus granulation, causes and management, Intensive Agriculture (November-December) pp.6-7.

Venkatesan, K., Balakrishnan, K and Sambandamurthi, S. 1994. Leaf sheath loosening in banana. The Hindu, November 9, P. 8.

Verma P 2009. Physiological Disorders of Vegetable Crops. Alfa Beta Technical Solutions. Jaipur-India, p. 170.

Vijayalakshmi, D. and Srinivasan, P.S. 1999. Physiological maladies and remedies in grapes. Kissan World (July) pp. 57-59.

Vijayaraghavan, H., Jagannathan, R. and Wilfred Manuel, W. 2001. Pencil point disorders in coconut. The Hindu February 22. pp 1-2.

Wainwright, H., and Burbage. 1989. Physiological disorders in mango J. Hort. Sci. 64:125-135.

Wang Chien Yi. 1994. Chilling injury of tropical horticultural commodities. Hort Sci 29: 986-988.

Weller, S.C. and Hammer, P.A. 1984. Susceptibility of Easter lily to glyphosate injury. Hort. sci 19: (5) 698-699.

Wien, H. 1997. Physiology of vegetable crops. CABI, USA p.672.

Wittwer, S.H. 2009. Foliar application of nutrients - Part of the chemical revolution in agriculture, Plant Food Review 2, National Plant Food Institute, Washington DC.

Zhang, C.L., Huang, H.B., and Kuang, Y.H. 1995. A study of the cause of the mango black tip disorder. Sci. Hort. 64:49-54.

□□□

Appendices

1. Conversion Tables

a) Length

Centimetres (cm)	Cm or inches	Inches
2.54	1	0.394
5.08	2	0.787
7.62	3	1.181
10.16	4	1.575
12.70	5	1.969
15.24	6	2.362
17.78	7	2.756
20.32	8	3.150
22.86	9	3.543
25.40	10	3.937
50.80	20	7.874
76.20	30	11.811
101.60	40	15.748
127.00	50	19.685

Kilometres (km)	Km or miles	Miles
1.609	1	0.621
3.219	2	1.243
4.828	3	1.864
6.437	4	2.485
8.047	5	3.107
9.656	6	3.728
11.265	7	4.350
12.875	8	4.971

14.484	9	5.592
16.093	10	6.214
32.187	20	12.427
48.280	30	18.641
64.374	40	24.855
80.467	50	31.069

b) Area

Hectares (ha)	Ha or acres	Acres
0.405	1	2.471
0.809	2	4.942
1.214	3	7.413
1.619	4	9.884
2.023	5	12.355
2.428	6	14.826
2.833	7	17.297
3.327	8	19.769
3.642	9	22.240
4.047	10	24.711
8.094	20	49.421
12.140	30	74.132
16.187	40	98.842
20.234	50	123.553

c) Mass (Weight)

Kilograms (kg)	Kg or lb	Pounds (lb)
0.454	1	2.205
0.907	2	4.409
1.361	3	6.614
1.814	4	8.819
2.268	5	11.023
2.722	6	13.228
3.175	7	15.432
3.629	8	17.637
4.082	9	19.842
4.536	10	22.046

9.072	20	44.092
13.608	30	66.139
18.144	40	88.185
22.680	50	110.231

d) Volume

Litres	Litres or gallons	Gallons
4.546	1	0.220
9.092	2	0.440
13.628	3	0.660
18.184	4	0.880
22.730	5	1.100
27.276	6	1.320
31.822	7	1.540
36.368	8	1.760
40.914	9	1.980
45.460	10	2.200
90.919	20	4.399
136.379	30	6.599
181.839	40	8.799
227.298	50	10.988

2. Symbols and relative atomic masses ("atomic weights") of some elements

Symbol	Element	Relative atomic mass
Al	aluminium	27
B	boron	10.8
C	carbon	12
Ca	calcium	40.1
Cl	chlorine	35.5
Co	cobalt	58.9
Cu	copper	63.5
Fe	iron	55.9
H	hydrogen	1
K	potassium	39.1
Mg	magnesium	24.3
Mn	manganese	54.9

Mo	molybdenum	95.9
N	nitrogen	14
Na	sodium	23
O	oxygen	16
P	phosphorus	31
S	sulphur	32.1
Si	silicon	28.1
Zn	zinc	65.4

3. Units

Mt	million metric tons (10^3 kt)
Kt	thousand metric tons (10^3 t)
t	metric ton (10^3 kg)
kg	kilogram (10^3 g)
g	gram (10^{-3} kg)
mg	milligram (10^{-3} g)
μg	microgram (10^{-3} mg)
ng	nanogram (10^{-3} μg)
pg	picogram (10^{-3} ng)
fg	femtogram (10^{-3} pg)
ppm	μg/g = mg/kg =g/t
ppb	10^{-3} ppm = ng/g = μg/kg = mg/t
ha	hectare (10,000 m^2)
L	litre (1 dm^3)
mL	millilitre; 10^{-3} L
m^3	cubic metre (10^3 dm^3).

4. Soil reaction pH

Degree of acidity or alkalinity	pH
Extremely acid	< 4.5
Very strongly acid	4.5-5.0
Strongly acid	5.1-5.5
Medium acid	5.6-6.0
Slightly acid	6.1-6.5
Neutral	6.6-7.3
Mildly alkaline	7.4-7.8

Moderately alkaline	7.9-8.4
Strongly alkaline	8.5-9.0
Very strongly alkaline	9.1 and above

5. Relative sensitivity of selected horticultural crops to micronutrient deficiencies

Crop	Sensitivity to micronutrient deficiency					
	B	Cu	Fe	Mn	Mo	Zn
Bean	Low	Low	Medium	Medium	Medium	Low
Broccoli	Medium	Medium	High	Medium	High	-
Cabbage	Medium	Medium	Medium	Medium	Medium	-
Carrot	Medium	Medium	-	Medium	Low	Low
Cauliflower	High	Medium	High	Medium	High	-
Celery	High	Medium	-	Medium	Low	-
Cucumber	Low	Medium	-	Medium	-	-
Lettuce	Medium	High	-	High	High	Medium
Onion	Low	High	-	High	High	Medium
Pea	Low	Low	-	High	Medium	Low
Radish	Medium	Medium	-	High	Medium	-
Spinach	Medium	High	High	High	High	High
Table beet	High	High	High	High	High	Medium
Tomato	Medium	Medium	High	Medium	Medium	Medium
Turnip	High	Medium	-	Medium	Medium	-

6. Safe ranges of macro and micronutrients application in vegetable crops

Nutrients	Compounds	Soil application (kg/ha)	Foliar application (%)
Nitrogen	Urea	75-50	2.0-3.0
Iron	Ferrous sulphate	25-50	0.2-0.5
Manganese	Manganese sulphate	25-50	0.2-0.5
Zinc	Zinc sulphate	25-50	0.2-0.5
Boron	Sodium tetraborate	15-20	0.3-0.4
Molybdenum	Sodium or Ammonium molybdate	0.1-1.0	0.02
Copper	Copper sulphate	10-20	0.2-0.5

7. Planting Distance Calculation

a) Square system of planting

$$\frac{\text{Area of the land}}{\text{Area occupied by a single tree}}$$

Area of the land = 1 ha. (10000 sq.m^2)

Spacing between the plants and rows = 10 m

Area occupied by a single tree = 10 m x 10 m = 100 m^2

No. of plants required per hectare $= \frac{10,000\ m^2}{100\ m^2} = 100$ plants

b) Rectangular system

$$\frac{\text{Area of the land}}{\text{Area occupied by a single tree}}$$

Area of the land = 1 ha. (10000 sq.m^2)

Spacing between the plants = 10 m

Spacing between the rows = 12 m

Area occupied by a single tree = 10 m x12 m = 120 m^2

No. of plants required per hectare $= \frac{10,000\ m^2}{120\ m^2} = 88$ plants

c) Quincunx system

$$\frac{\text{Area of the land}}{\text{Area occupied by a single tree}} \text{ X } 2$$

(Double the no. of plants of a square system)

Area of the land = 1 ha. (10000 sq.m^2)

Spacing between the plants and rows = 10 m

Area occupied by a single tree = 10 m x 10 m = 100 m^2

No. of plants required per hectare

$$= \frac{10,000\ m^2}{100\ m^2} = 100 \text{x} 2 = 200 \text{ plants}$$

As the plants are planted additionally in the centre of the square, hence first, the no. of plants are calculated for square system of planting which is:

$$\frac{\text{Area of the land}}{\text{Area occupied by a single tree}}$$

$$= \frac{10{,}000 \text{ m}^2}{10 \text{ x } 10 \text{ m}} = 100$$

Additional plants = (No. of rows length wise -1) X (No. of rows width wise -1)

In 100 x 100 Sq metre field, if planting distance 10 x 10 m, then the number of rows length wise and width wise will be 10.

Hence, no. of plants = (10-1) x (10-1) = 9x9 = 81.

So, total no. of plants = plants planted in square system of planting + additionally planted plants in the center of the square = 100 + 81 = 181.

d) Hexagonal system

Area occupied by a single tree $\frac{\text{Area of the land}}{\text{Area occupied by a single tree}}$

Method-I

In this system, each tree occupies a parallelogram area

Each parallelogram consists of 2 equilateral triangles.

Area of each parallelogram consists of 2 equilateral triangles

Area of each equilateral trianglses is $\frac{\sqrt{3}}{4} x\, a^2\, x\, 2$

Where a is the length of a side of a triangle i.e. Spacing between the trees in a row.

So, a tree occupies $= \frac{\sqrt{3}}{4} x\, a^2\, x\, 2$

Suppose if spacing is 10 m i.e. a

Area occupied by a single tree is $\frac{\sqrt{3}}{4} x\, a^2\, x\, 2$

= 1.732 X 10 X 10 X 2 = 86.60 m^2

Area of the land = 10000 m^2

No. of plants per hectare $= \frac{10{,}000 \text{m}^2}{86.60 \text{ m}^2} = 115/116$

e) Number of plants/acre based on spacing

S.No.	Spacing	No. of plants/acre
1.	5' x 5'	1,742
2.	6' x 6'	1,210
3.	7' x 7'	869
4.	8' x 8'	680
5.	9' x 9'	537
6.	10' x 10'	435
7.	11' x 11'	360
8.	12' x 12'	302
9.	13' x 13'	257
10.	14' x 14'	222
11.	15' x 15'	193
12.	16' x 16'	170
13.	17' x 17'	150
14.	18' x 18'	134
15.	19' x 19'	120
16.	20' x 20'	108
17.	22' x 22'	90
18.	24' x 24'	75
19.	26' x 26'	64
20.	28' x 28'	55
21.	30' x 30'	48
22.	40' x 40'	27

8. Nutrients for Foliar spray in citrus

Application of the following nutrients as foliar spray has been suggested after the new leaves have expanded to at least 2/3 rd of their normal size (Citrus).

Zinc sulphate	2.5 kg
Copper sulphate	1.5 kg
Manganese sulphate	1.0 kg
Magnesium sulphate	1.0 kg
Ferrous sulphate	1.0 kg
Borax or boric acid	0.5 kg
Urea	5.0 kg
Lime	4.5 kg
Water	450 litre

9. Fertilizer efficiencies by various application methods

Nutrient	Fertilizer use efficiency (%)	
	Soil application	Fertigation
Nitrogen	30-50	95
Phosphorous	20	45
Potassium	50	80

10. Lime Requirement for Acidic Soils

p^H of soil buffer suspension (Soil taken from the field)	Lime required to bring pH down to indicated level (Calcium carbonate) in tonnes / acre		
	p^H 6.0	p^H 6.4	p^H 6.8
6.7	1.0	1.2	1.4
6.6	1.4	1.7	1.9
6.5	1.8	2.2	2.5
6.4	2.3	2.7	3.1
6.3	2.7	3.2	3.7
6.2	3.1	3.7	4.2
6.1	3.5	4.2	4.8
6.0	3.9	4.7	5.4
5.9	4.4	5.2	6.0
5.8	4.8	5.7	6.5
5.7	5.2	6.2	7.1
5.6	5.6	6.7	7.7
5.5	6.0	7.2	8.3
5.4	6.5	7.7	8.9
5.3	6.9	8.2	9.4
5.2	7.4	8.4	10.0
5.1	7.8	9.1	10.6
5.0	8.2	9.6	11.2
4.9	8.6	10.1	11.8
4.8	9.1	10.6	12.4

11. Indicator plant species for various nutrient deficiencies

Deficient element	Indicator plants
Nitrogen	Maize, non-leguminous small grains, mustard, apple, citrus
Phosphorus	Maize, barley, lettuce, tomato
Potassium	Potato, clover, Lucerne, bean, tobacco, cucurbits, cotton, tomato, maize
Calcium	Lucerne, other legumes
Magnesium	Potato, cauliflower
Sulphur	Lucerne, clover
Iron	Sorghum, barley, citrus, peach, rice
Copper	Apple, citrus, barley, maize, lettuce, oats, onion, tobacco, tomato
Manganese	Apple, apricot, bean, cherry, citrus, maize, oats, pea, radish, wheat
Boron	Lucerne, turnip, cauliflower, apple, peach
Molybdenum	Cauliflower, other Brassica spp. Citrus, legumes, oats, spinach
Chlorine	Lettuce

12. Critical concentration of nutrients in soil and plants

a) Major Nutrients

Nutrients	Plants	Soil Limits	
N	1.5%	Below 250 kg/ha 250-500 kg/ha Above 500 kg/ha	Low Medium High
P	Below 0.2%	Below 10 kg P/ha 10-20 kg P/ha Above 20 kg P/ha	Low Medium High
K	1.4	Below 100 kg/ha 100-250 kg/ha Above 250 kg/ha	Low Medium High

b) Secondary and micro nutrients

Nutrients	Plants	Soil Limits
S	Below 0.15-0.2%	Below 10 ppm hot water soluble
Ca	Below 0.2%	Below 50% of CEC Ammo. Acetate Extractable
Mg	0.1 to 0.2%	Below 4% of CEC Ammo. Acetate Extractable
Zn	Below 20 ppm (15-20)	Below 0.6 ppm (0.5 to 1.2ppm) DTPA Extractable
Fe	Below 50 ppm (25-80)	2.5-4.5 ppm DTPA Extractable below 2ppm Ammo. Acetate Extractable
Mn	Below 20 ppm (20-30)	Below 2ppm DTPA Extractable
Cu	Below 4 ppm (3-10)	Below 0.2 ppm DTPA Extractable
B	Below 1 ppm	Below 0.5 ppm hot water soluble
Mo	Below 0.1 ppm	Below 0.2 ppm Ammo. Oxalate extractants
Cl	Below 100 ppm	Below 2.0 ppm water soluble

13. Nutrients Supplied by Fertilizers

Nutrient	Fertilizers
1 kg Nitrogen	= 5 kg Ammonium sulphate= 4 kg Calcium ammonium nitrate= 4 kg Ammonium chloride= 2.2 kg Urea= 3.3 kg Ammonium nitrate= 5.5 kg Di ammonium phosphate
1 kg Phosphorus	= 6.25 kg Super phosphate= 2.02 kg Di ammonium phosphate
1 kg Potassium	= 1.660 kg of Muriate of Potash = 2.000 kg of Sulphate of Potash
1 kg Calcium	= 2.5 kg Lime stone = 5.0 kg Dolomite = 4.5 kg Gypsum
1 kg Magnesium	= 10.4 kg Magnesium Sulphate - $7H_2O$
1 kg S	= 8.3 kg Single super Phosphate= 4.2 kg Ammonium sulphate= 5.5 kg Gypsum
1 kg B	= 8.8 kg Borax = 4.8 kg SOLUBOR
1 kg Cu	= 4.2 kg Copper sulphate
1 kg Fe	= 5.3 kg Ferrous sulphate
1 kg Mn	= 3.3 kg Manganese sulphate
1 kg Zn	= 4.8 kg Zinc sulphate
1 kg Mo	= 1.92 kg Ammonium molybdate

14. Interactions between nutrients in Plants

Major Element	Antagonistic Elements	Synergistic Elements
Ca	Al, B, Fe, Mn, Ni and Zn	Cu, Mn and Zn
Mg	Al, Mn, Zn, Cu and Fe	Al and Zn
P	Cu, Mo, Mn, Ni, Se, Si and Zn	Al, B, Cu, F, Fe, Mo, Mn and Zn
K	B, Mo, Mn	-
S	As, Ba, Fe, Mo, Pb, Se and Zn	Fe
N	B, Cu and Mn	B, Cu, Fe and Mo
Cl	Br and I	-
Na	Mn	-
Si	B, Mn	-

15. Nutrient content in Plants

Nutrient	Low	Marginal	Sufficient	High	Excess
Nitrogen (N) %	1.5	1.5 - 2.0	2.0 - 3.0	3.0 - 4.0	4.0
Phosphorous (P) %	0.15	0.15 - 0.25	0.26 - 0.5	0.5 - 0.8	0.8
Potassium (K) %	1.0	1.0 - 1.5	1.5 - 3.0	3.0 - 5.0	5.0
Sulphur (S) %	0.1	0.1 - 0.15	0.15 - 0.40	0.40 - 0.8	0.8
Calcium (Ca) %	0.10	0.10 - 0.2	0.2 - 1.0	1.0 - 1.5	1.5
Magnesium (Mg) %	0.1	0.1 - 0.15	0.15 - 0.50	0.5 - 1.0	1.0
Zinc (Zn) ppm	10	10 - 15	15 - 70	70 - 150	150
Copper (Cu) ppm	2.3	2.3 - 3.7	3.7 - 25	25 - 50	50
Iron (Fe) ppm	15	15 - 20	20 - 250	250 - 500	500
Manganese (Mn) ppm	10	10 - 15	15 - 100	100 - 250	250
Boron (B) ppm	3	3 - 5	5 - 25	25 - 75	75
Molybdenum (Mo) ppm	0.01	.01 - .02	.03 - 5	5 - 10	10

16. Ready reckoner for the preparation of spray fluid

Parts per Million	Percentage	Necessary chemical g/10 litres
1	0.0001	0.01
10	0.001	0.10
25	0.0025	0.25
50	0.005	0.50
75	0.0075	0.75
100	0.01	1.00
200	0.02	2.00
300	0.03	3.00
400	0.04	4.00
500	0.05	5.00
600	0.06	6.00
700	0.07	7.00
800	0.08	8.00
900	0.09	9.00
1000	0.010	10.00
1250	0.125	12.50
1500	0.150	15.00
1750	0.175	17.50
2000	0.200	20.00

17. Preparation of solution of concentration from given stock solution

(S) Stock solution (ppm)	(R) Required concentration (ppm)	(S/R) Dilute ml of stock solution to volume (ml)
10,00,000	1,000	1,000
1,00,00	1,000	100
10,000	100	100
1,000	100	10
100	10	10
10	1	10

18. Equivalent solution

Equivalent dilutions (1g = 1000 mg = 10,00,000 μg)

%	gm/1itre	mg/1itre	ppm/1itre	mg/g
10	100	1,00,000	1,00,000	1,00,000
1	10	10,000	10,000	10,000
0.1	1	1,000	1,000	1,000
0.01	0.1	100	100	100
0.005	0.05	50	50	50
0.001	0.01	10	10	10
0.0001	0.001	1	1	1

20. Recommendation of Micronutrient fertilizers related to their availability

a) Recommendation of Zn fertilizers related to available Zn content of soil

Available Zn in soil (mg/kg	Zn fertility level	Recommendations for Zn application in crops
< 0.3	Acute deficiency	5.0 kg Zn/ha basal (25 kg $ZnSO_4$. $7H_2O$/ha) or 15 kg $ZnSO_4$. H_2O/ha per crop sequence
0.3-0.6	Deficiency	5.0 kg Zn/ha basal (25 kg $ZnSO_4$. $7H_2O$/ha) or 15 kg $ZnSO_4$. H_2O in every second year crop or 3.0 kg Zn/ha basal (15 kg $ZnSO_4$. $7H_2O$/ha) or 15 kg $ZnSO_4$. H_2O/ha per crop sequence
0.6-0.9	Marginal deficiency	5.0 kg Zn/ha basal (25 kg $ZnSO_4$. $7H_2O$/ha) or 15 kg $ZnSO_4$. H_2O /ha every third year crop sequence or 2.5 kg Zn/ha basal (12.5 kg $ZnSO_4$.$7H_2O$/ ha) or 7.5 kg $ZnSO_4$. H_2O/ha per crop sequence
0.9-1.2	Partially adequate	2.0 kg Zn/ha basal per crop sequence
1.2-1.8	Adequate	Maintenance dose of 1.5 kg Zn/ha basal (7.5 kg $ZnSO_4$.$7H_2O$) or 4.5 kg $ZnSO_4$. H_2O per crop sequence
> 1.8	High	No need of Zn application

b) Recommendations for B application related to available B content of soil on the basis of available B content in soil.

Available B in soil (mg/kg^{-1})	B fertility level	Recommendations for B application in crops
< 0.3	Acute deficiency	2.0 kg B per year in ever crop sequence
0.3-0.5	Deficiency	1.5 kg B per year in ever crop sequence
0.5-0.7	Marginal deficiency	1.0 kg B per year in ever crop sequence
0.7-0.9	Partially adequate	Maintenance dose of 0.5kg B ha^{-1} in every crop
0.9-1.2	Adequate	No need of B application
>1.2	High	No need of B application

c) Recommendations for Fe application related to available Fe content of soil

Available Fe in soil (mg/kg)	Fe fertility level	Recommendations of Fe application in crops
< 3.5	Acute deficiency	50 kg $FeSO_4$ basal per crop sequence + 1 foliar spray of 1% $FeSO_4$ at the time of maximum tillering/ branching.
3.5-5.5	Deficiency	40 kg $FeSO_4$ basal per crop sequence + 1 foliar spray of 1% $FeSO_4$ at the time of maximum tillering/ branching.
5.5-7.5	Marginal deficiency	30 kg $FeSO_4$ basal per crop sequence or 3 foliar sprays of 1% $FeSO_4$ at the time of critical growth stages.
7.5-9.5	Partially adequate	2 foliar sprays of 1% $FeSO_4$ at the time of critical growth stages.
9.5-11.5	Adequate	1 foliar spray of 1% $FeSO_4$ at the time of maximum

		tillering/branching for enhancing Fe content in edible plant part otherwise no need of application.
>11.5	High	No need of Fe application

d) Recommendations for Mn application related to available Mn content of soil

Available Fe in soil (mg/kg)	**Fe fertility level**	**Recommendations of Fe application in crops**
< 2.0	Acute deficiency	50 kg $MnSO_4$ basal per crop sequence or 20 kg $MnSO_4$ basal per crop sequence + 3 foliar spray of 0.5% $MnSO_4$ at the time of critical growth stages.
2.0-4.0	Deficiency	40 kg $MnSO_4$ basal per crop sequence or 20 kg $MnSO_4$ basal per crop sequence + 2 foliar spray of 0.5% $MnSO_4$ at the time of grand growth stages.
4.0-6.0	Marginal deficiency	30 kg $MnSO_4$ basal per crop sequence or 20 kg $MnSO_4$ basal per crop sequence + 1 foliar spray of 0.5% $MnSO_4$ at the time of grand growth.
6.0-8.0	Partially adequate	2 foliar spray of 0.5% $MnSO_4$ at the time of grand growth stages.
8.0-10.0	Adequate	1 foliar spray of 0.5% $MnSO_4$ as per the crop demand
> 10.0	High	No need of Mn application

20. Micronutrient content (ppm)

Organic material	Fe	Mn	Cu	Zn	B
a) Animal manures					
Cattle manure	690-1518	167-389	24-40	128-183	13-30
Farm Yard manure (FYM)	1135-3515	229-668	17-29	83-128	12-27
Goat manure	1449-2174	246-505	20-38	112-184	29-66
Pig manure	1020-1990	180-207	45-48	186-575	4-13
Poultry manure					
Broiler 1-2 litter	723-1565	213-421	27-40	166-271	15-27
Layer litter	1144-2215	287-450	22-38	182-329	12-23
b) Other dead materials					
Biogas residue	1097-1644	402-616	37-53	300-412	22-46
Compost	2090-4064	201-505	11-25	75-169	12-20
Sewage sludge	2275-3322	100-287	7-11	69-177	3-5
c) Green manures	1020-1990	180-207	45-48	186-575	4-13
Acacia	870-994	78-92	7-9	54-61	46-49
Calapogonium	340-542	80-140	4-6	64-69	30-33
Centrosema	252-420	38-62	4-7	42-49	79-84
Gliricidia	212-450	85-156	4-5	57-70	41-69
Pueraria	117-280	82-120	4-6	64-70	20-24

21. Compatibility of Fertilizers

	1	2	3	4	5	6	7	8	9	10	11	12
1. Muriate of Potash	R	R	R	O	O	O	O	O	R	R	R	R
2. Sulphate of Potash	R	R	R	O	O	O	R	O	R	R	R	R
3. Sulphate of Ammonia	R	R	R	R	O	O	X	O	R	R	X	R
4. Calcium Ammonium Nitrate	O	O	R	R	O	O	X	O	O	O	X	O
5. Potassium & Sodium Nitrate	O	O	O	O	R	O	O	O	O	O	R	O
6. Calcium Nitrate	O	O	O	O	O	R	O	O	X	X	O	X

7.	Calcium Urea Cyanamide	O	R	X	X	O	O	R	O	X	X	R	X
8.	Urea	O	O	O	O	O	O	O	O	O	O	O	O
9.	Super Phosphate	R	R	R	O	O	X	X	O	R	R	X	R
10.	Ammonium Phosphate	R	R	R	O	O	X	X	O	R	R	X	R
11.	Basic Slag	R	R	X	X	R	O	R	O	X	X	R	X
12.	Diammonium Phosphate	R	R	R	O	O	X	X	O	R	R	X	R

R: Fertilizers can be mixed. O: Fertilizers may be mixed before use. X: Fertilizers cannot be mixed.

22. Classification of fruit crops according to soil pH

Slightly tolerant to acid soil pH 6.0-6.8	Moderately tolerant to acid soil pH 5.5-6.8	Highly tolerant to acid soil pH 5.5-6.8 and less	Tolerant to alkaline soil pH 7.2-8.0 and above
Mango	Pineapple	Strawberry	Custard apple
Citrus	Orange	Gooseberry	Guava
Banana	Litchi	Raspberry	Date
Guava	Longan	Blueberry	Coconut
Papaya	Avocado	Fig	Aonla
Cashew	Passion fruit	Bael	Olive
Carambola	Chestnut	Elephant apple	Phalsa
Apple	Cranberry	Plum	
Peach	Loquat		
Almond	Jack fruit		
Kiwi fruit			

23. Salinity of irrigation water

i) STL classification (Based on EC)

Good	:	<1 dSm^{-1}
Critical	:	1-3 dSm^{-1}
Injurious	:	>3 dSm^{-1}

ii) USSL Classification

EC (dSm^{-1}) at 25°C	Water class	Interpretation
< 0.25	Low salinity (C_1)	Safe with no likelihood of any salinity problem developing
0.25-0.75	Medium salinity (C_2)	Need moderate leaching
0.75-2.25	High salinity (C_3)	Cannot be used on soils with inadequate drainage since saline conditions are likely to develop.
> 2.25	Very high salinity (C_4)	Cannot be used on soils with inadequate drainage since saline conditions are likely to develop.

2. Alkalinity (Based on pH)

pH	Classification	Interpretation
6.0-8.0	Normal	Can be used on all soils
8.1-8.5	Medium alkaline	Can be used on soil with good drainage and in the presence of enough organic matter
8.6 and above	Alkaline	In addition to all the criteria needed for the use of medium alkaline water, certain special measures like application of gypsum, superphosphate, etc., should be taken up.

3. Sodium Adsorption Ratio (SAR)

SAR is a parameter calculated based on Ca^{2+}, Mg^{2+} and Na^{+} concentration of irrigation water to know the alkalinity hazard of irrigation water as given below:

$$SAR = \frac{Na}{[Ca + Mg/2]^{1/2}}$$

where, concentration of cations is in me L^{-1}.

Low : < 10

Medium : 10.1-18.0

High : 18.1-26.0

Very high : >26

4. Residual Sodium Carbonate (RSC)

This index is important for carbonate and bicarbonate rich irrigation waters. It indicates their tendency to precipitate Ca^{2+} as $CaCO_3$.

$RSC = (CO_3^{2-} + HCO_3^{-}) - (Ca^{2+} + Mg^{2+})$

Good : < 1.25 me L^{-1}

Fair : 1.25-2.5 me L^{-1}

Unsuitable : > 2.5 me L^{-1}

5. Soluble Sodium Percentage (SSP)

$$SAR = \frac{Na}{Ca^2 + Mg^2 + K^+ + Na^+} x100$$

Unsuitable : > 60%

6. Potential Salinity (PS)

$PS = ½ SO_4^{2-} + Cl^-$

Sandy soil : 5-20 me L^{-1}

Medium texture : 3-15 me L^{-1}

Fine texture : 1-3 me L^{-1}

7. Puri's Salt Index (PSI)

PSI = (Total Na^+ - 24.5) - (Total Ca^{2+} Ca as $CaCO_3$) x 4.85

Good : -ve

Poor : +ve

Note : The concentration of cations and anions should be expressed in me L^{-1} for deriving the above quality parameters.

8. Chloride concentration

Chloride (me L^{-1})	Class
0-5	Excellent for irrigation
5.1-1	Good
0> 10	Injurious to crops

9. Boron concentration

Low	:	< 0.5 ppm
Medium	:	0.5-1.0 ppm
High	:	1.0-2.0 ppm
Very high	:	2.0-4.0 ppm
Excessive	:	>4.0 ppm

10. Fluorine concentration

Good	:	< 1 ppm
Problematic	:	> 1 ppm

11. Combined evaluation on the basis of EC & SAR

United States Salinity Laboratory published a semi logarithmic diagram, which rates water with respect to EC and SAR. Four classes of each salinity and SAR were divided, as shown below in the table.

Salinity Hazard	Class	EC (dS m^{-1}) at 25°C	Sodium Hazard	Class	SAR
Low	C1	< 0.25	Low	S1	< 10
Medium	C2	0.25-0.75	Medium	S2	10-18
High	C3	0.75-2.25	High	S3	18-26
Very high	C4	> 2.25	Very high	S4	> 26

24. Classification of fruits based on relative salt tolerance

High tolerance	Medium tolerance	Highly sensitive
Date palm	Pomegranate	Apple
Ber	Jamun	Pear
Aonla	Fig	Mango
Bael	Olive	Citrus
Guava	Longan	Strawberry
Coconut	Passion fruit	Avocado
Khirni	Phalsa	Durian
	Cashewnut	

25. Tips for effective foliar nutrition

- Have the deficiency symptoms or disorder diagnosed by local specialist/adviser is checked before spraying.
- Do not exceed the recommended solution concentration.
- Adjust concentration according to type of spray used (Particle size of the spray droplets)
- Do not spray during hot, windy periods and against the breeze.
- Wherever necessary, neutralize spray solution with lime.
- Use only clear filtered solution
- Repeat spray as and when recommended.
- Avoid spray during misting to prevent washing away of the spray.
- Double check the calculations used for preparing the final solution.
- Do not expect a foliar spray to leave residual effect tor next crop.
- Do not store spray solution, particularly of iron. Use freshly made solutions.
- Use wetting agents to improve effectiveness of spray @ 1 ml/ litre of spray fluid.
- Before buying a product, examine clearly its nutrient makeup and composition and whether it meets the growing demand.

Precautions for foliar fertilization

In hardness or amount of salts contained in the water used to make the solutions affects the degree of concentration of which fertilizer chemicals may be applied with safety. The 'harder' the water, the greater the concentration of the fertilizer chemicals can be used. Hydrated lime can be used in soft water to enable the absorption more concentrated solutions.

26. Tips for high fertilizer use efficiency

- Priority should be given to crops which have controlled irrigation as compared to unirrigated crops.

- Most responsive and best suited crop variety should be selected to get maximum benefits of fertilizer.
- High yielding varieties give better yield response to fertilizers than local varieties. Therefore, high-yielding varieties should be preferred.
- Planting/sowing of crop should be done at the normal time suited for the tract to get maximum benefit from applied fertilizers.
- Proper plant spacing has to be ensured
- It is desirable to add organic waste and organic manure to built soil fertility and its water holding capacity of soil
- Leguminous crops are taken in rotational and as inter crop and use of bacterial and algae cultures play important role in meeting the fertilizer need of the crop
- Irrigation rate should be controlled by using just enough water at different physiological growth stages of crops
- Balanced fertilizer should be practiced based on the soil test
- Fertilizer recommendation should preferably be based on crop sequence for multiple cropping
- All the phosphate and potash should be applied as basal dose, while nitrogen should be applied in split doses. (2-3 times)
- Phosphate should be placed 4 to 6 cm, below and away from the seed to ensure maximum availability to the crop potash should be broad casted then mixed with the soil
- Urea must be applied with soil for top dressing to reduce nitrogen loss by mixing 1 part of urea with 5-10 parts of moist soil and keep it for 24 hours for getting best results.
- Under adverse soil and climate conditions, application of fertilizers, such as nitrogen through foliage along with insecticides and pesticides will lead to higher utilization by plants
- When soil is acidic, saline and alkaline appropriate amendments viz. Lime, Gypsum etc. should be applied before using fertilizers. In alkali soils 3 to 5 tonnes of gypsum per acre should be applied.
- Weed control, particularly during early stages of crop growth is essential
- Proper control of pests and diseases is a must for realizing maximum effectiveness from fertilizers.

27. Foliar application of nutrients

a) Fruits

S.No.	Crop	Nutrients and doses	Time of application	Remarks
1.	Mango	2% K_2SO_4	Two sprays at pea stage and 15 days after	To improve yield and quality
2.	Banana	0.5% $ZnSO_4$ + 0.2% $FeSO_4$ + 0.2% $CuSO_4$+ 0.1% H_3BO_3	3rd, 5th and 7th months after planting	To enhance yield and quality.
3.	Acid lime	0.5% $ZnSO_4$	Three times viz., March, July and October	To increase flowering
4.	Grapes	1.0% urea + 0.2% $ZnSO_4$+ 0.1% H_3BO_3	Twice, before flowering and 10 days there after	To enhance the yield and quality
5.	Guava	0.3% H_3BO_3	Flowering and fruit set	To enhance the size and reduces the cracking
6.	Pineapple	0.5% $ZnSO_4$ + 0.5% $FeSO_4$	Twice at 15 days interval during vegetative stage	To enhance the growth and yield
7.	Papaya	0.5% $ZnSO_4$ + 0.1% H_3BO_3	4th and 8th months after planting	To enhance growth and yield

b) Vegetables

S.No.	Crop	Nutrients and doses	Time of application	Remarks
1.	Tomato	0.5% $ZnSO_4$	Thrice at 10 days interval from 40 days after planting.	To enhance flowering and yield.
2.	Okra	1.0% urea + 1.0% KCl	30 and 45 days after planting.	To enhance the yield.

c) Flowers

S.No.	Crop	Nutrients and doses	Time of application	Remarks
1.	Rose	0.2% of the mixture containing 20g $MnSO_4$ + 15g $MgSO_4$ + 10 g $FeSO_4$+ 5g Boric acid	Flower bud initiation	To enhance the colour of the flowers
2.	Jasmine (Malligai)	0.5% $MgSO_4$ + 0.5% $FeSO_4$ + 0.25% $ZnSO_4$	Two sprays at 15 days interval at vegetative stage.	To enhance the growth and yield.
3.	Chrysan themum	0.5% $MgSO_4$ + 0.25% $ZnSO_4$	Twice at an interval of 15 days.	To increase yield
4.	Marigold	0.5% $FeSO_4$ + 0.5% $ZnSO_4$ + 0.2% Humic acid	30 and 45 days after transplanting	Increase the yield.
5.	Tuberose	0.5% $ZnSO_4$ + 0.2% $FeSO_4$ + 0.1% H_3BO_3 yield.	45 and 60 days after planting	To boost growth and

28. Application of plant growth regulators

a) Fruits

S.No.	Crops	Name and doses	Time of application	Remarks
1.	Mango	(i) NAA @ 20ppm (ii) Paclobutrazol @ .75g/meter canopy radius	Flowering 90 days prior to flowering	Increases fruit set. To induce flowering
2.	Banana	Plantozyme @ 2 ml/litre	6th and 8th month after planting	Enhanced yield
3.	Acid lime	NAA 30ppm	After fruit set	Increased yield

4.	Grapes	GA_3 25ppm	10-12 days after fruit set	Higher yield
5.	Pineapple	NAA 200ppm	After fruit formation	Increases size of the fruit
6.	Papaya	GA_3 100ppm	Soak the seeds for 16 hours	Germination enhanced.

b) Vegetables

S.No.	Crops	Name and doses	Time of application	Remarks
1.	Tomato	Triacontanol 1.25 ppm	15 days after transplanting and full bloom stage.	Increases yield.
2.	Chillies	NAA 20ppm	60 and 90 days after transpl-anting.	Increases yield.
3.	Pumpkin	Ethrel 250ppm	Four times (weekly inter-vals) from 10-15 days\after sowing.	Increases yield.

c) Flowers

S.No.	Crop	Nutrients and doses	Time of application	Remarks
1.	Rose	GA_3@ 250 ppm	30 days after pruning.	Increase flower production.
2.	Crossandra	Ascorbic acid 1000 ppm	Before flowering	Increase flower production.
3.	Chrysan themum	GA_3@ 50ppm	30, 45 and 60 days after planting.	Increase flower production.
4.	Tuberose	GA_3@ 100 ppm	40, 55 and 60 days after planting.	Increase flower production.
5.	Dendrobium orchid	GA_3 50 ppm	Bimonthly interval from 30 days after planting	Enhance flower production

29. Conversion of pure nutrients to various N, P and K fertilizers

Rate of application (kg/ha)	Ammonium sulphate (20% N)	Urea (46% N)	Super phosphate (18% P)	Muriate of potash	
				(50 % K)	(60 % K)
10	50	22	56	20	17
20	100	44	112	40	34
30	150	66	168	60	51
40	200	88	224	80	68
50	250	110	280	100	85
60	300	132	336	120	102
70	350	154	392	140	119
80	400	176	448	160	136
90	450	198	504	180	153
100	500	200	560	200	170
110	550	242	616	220	187
120	600	264	672	240	204
130	650	286	728	260	221
140	700	308	784	280	238
150	750	330	840	300	255

30. Some important conversion factors

N x 1.286 = NH_4

N x 4.43 = NO_3

Organic carbon x 1.724 = Organic matter

P x 2.29 = P_2O_5

K x 1.20 = K_2O

Ca x 1.40 = CaO

Ca x 1.85 = $Ca(OH)_2$

S x 3 = SO_4

Acre x 2.471 = hectare = 2.24 x 10^6 kg soil (plough layer)

Hectare x 0.404686 = Acre

Lb $acre^{-1}$ x 1.21 = kg ha^{-1}

me L^{-1} x equivalent wt. = ppm

ppm = μg mL^{-1} = mg L^{-1}

ppm = mg kg^{-1}

mm per opening = 16 ÷ mesh per inch

31. Nutrient status of vermicomposts prepared by different species of earthworms

Parameter	Pheritima Elongata	Eisenia Foetida	Perionyx Excavator	FYM
pH	7.2	7.4	7.0	7.2
Electrical conductivity (dS/m)	0.38	0.90	1.20	0.18
		Major nutrients		
Total N (%)	0.42	0.60	0.66	0.5
Avail N(ppm)	215.0	450	496	375
Total P_2O_5 (%)	1.16	1.34	1.93	0.75
Total K_2O (%)	0.26	0.40	0.42	2.30
Available K_2O (ppm)	3000	4500	4700	26,030
		Micro nutrients		
Fe (ppm)	27.3	17.8	19.8	24.7
Zn (ppm)	18.0	19.2	0.9	40.0
Mn (ppm)	16.4	24.6	16.5	120
Cu (ppm)	7.6	7.6	2.3	2.8

32. Nutrient content of bulky organic manures

Manure	Percentage composition		
	N	P_2O_5	K_2O
Cattle dung	0.40	0.20	0.20
Cattle urine	1.00	0.01	1.35
Sheep and goat dung	0.75	0.50	0.45
Sheep and goat urine	1.35	0.05	2.10
Sheep and goat manure	3.00	1.00	2.00
Poultry manure	3.03	2.63	1.40
Horse manure	2.00	1.50	1.50
Horse urine	1.35	0.01	1.25
Pig dung	0.60	0.50	0.40
Pig urine	1.10	0.10	0.45
Farm litter compost	0.50	0.15	0.50

Rural compost	1.22	1.08	1.47
Town compost	1.40	1.00	1.40
Vermicompost	3.00	1.00	1.50
Night soil	5.50	4.00	2.00
Paddy straw	1.50	1.34	3.37
Sugarcane trash	2.73	1.81	1.31
Sewage sludge	1.5-3.5	0.75-4.00	0.3-0.6

33. Phytotoxic symptoms scoring and rating on weeds and crops

Weed control rating	Crop injury symptoms	Rating	Effect
No control	No injury, Normal	0	None
Very poor control	Slight stunting, injury or discoloration	1	Slight
Poor control	Some stand loss, stunting/discolouration	2	Slight
Poor to deficient control	Injury more pronounced but not persistent	3	Slight
Deficient control	Moderate injury, recovery possible	4	Moderate
Deficient to moderate control	Injury more persistent, recovery doubtful	5	Moderate
Moderate control	More severe injury, no recovery possible	6	Moderate
Satisfactory control	Severe injury, stand loss	7	Severe
Good control	Almost destroyed, few plants surviving	8	Severe
Good to excellent control	Very few plant alive	9	Severe
Complete control	Complete destruction	10	Complete

34. Calculation of Paclobutrazol

To induce flowering in mango, Paclobutrazol (Cultar) is recommended. Generally, it is applied 90 days prior to the flowering i.e. between 15th July to 31st August.

$$\text{Dose required} = \frac{\text{(N-S) M + (E-W) M}}{2} \times 3\,ml$$

Where

N-S refers to North - South canopy diameter

E-W refers to East - West canopy diameter

M = meter

3ml of Paclobutrazol (0.75g) at 23% active ingredient.

Method of application

After calculating the dose in ml, it is diluted with 6 litres of quality irrigation water and applied equally behind the fertilizer application zone ring by making 20 holes of 15-30 cm depth.

□□□